Encyclopaedia of

Tools and Techniques in Biochemistry and Molecular Biology

Encyclopaedia of

Tools and Techniques in Biochemistry and Molecular Biology

Volume 4

Editor:
Rashmi Rana

AGROTECH PRESS
JAIPUR ▪ NEW DELHI

AGROTECH PRESS

Regd. Off. "Bony Residency", Gate No. 2
Opp. Tilak Public School, Vishweswariya Nagar,
Gopalpura Road, Jaipur - 302018, India
Ph: 0141 - 2761280, Telefax: 2761381
E-mail : agrotechpress@gmail.com

Delhi Office: 102, 1st Floor, Satyam House,
4327/3, Ansari Road, Darya Ganj,
New Delhi - 110002, India
Ph: 011-45652440
visit us: www.agrotechpress.com

Distributed By
KEYA INTERNATIONAL PRESS LLP
E-mail: keyainternationalpressllp@gmail.com

Published in India, 2025

ISBN : 9789384568245 (4 Vols. Set)

Typeset by Thomson Press

Printed and bound in India by Thomson Press Ltd.,
New Delhi

Preface

Biochemistry, sometimes called biological chemistry, is the study of chemical processes in living organisms, including, but not limited to, living matter. The laws of biochemistry govern all living organisms and living processes. By controlling information flow through biochemical signalling and the flow of chemical energy through metabolism, biochemical processes give rise to the complexity of life. Much of biochemistry deals with the structures, functions and interactions of cellular components such as proteins, carbohydrates, lipids, nucleic acids and other biomolecules —although increasingly processes rather than individual molecules are the main focus. Among the vast number of different biomolecules, many are complex and large molecules, which are composed of similar repeating subunits. Each class of polymeric biomolecule has a different set of subunit types For example, a protein is a polymer whose subunits are selected from a set of 20 or more amino acids. Biochemistry studies the chemical properties of important biological molecules, like proteins, and in particular the chemistry of enzyme-catalyzed reactions. It once was generally believed that life and its materials had some essential property or substance distinct from any found in non-living matter, and it was thought that only living beings could produce the molecules of life Then, in 1828, Friedrich Wöhler published a paper on the synthesis of urea, proving that organic compounds can be created artificially.

The dawn of biochemistry may have been the discovery of the first enzyme, diastase (today called amylase), in 1833 by Anselme Payen. Eduard Buchner contributed the first demonstration of a complex biochemical process outside of a cell in 1896: alcoholic fermentation in cell extracts of yeast. Although the term "biochemistry" seems to have been first used in 1882, it is generally accepted that the formal coinage of biochemistry occurred in 1903 by Carl Neuberg, a German chemist. Previous to this time, this area would have been referred to as physiological chemistry. Since then, biochemistry has advanced, especially since the mid-20th century, with the development of new techniques such as chromatography, X-ray diffraction, dual polarisation interferometry,

NMR spectroscopy, radioisotopic labelling, electron microscopy, and molecular dynamics simulations. These techniques allowed for the discovery and detailed analysis of many molecules and metabolic pathways of the cell, such as glycolysis and the Krebs cycle (citric acid cycle). Researchers in biochemistry use specific techniques native to biochemistry, but increasingly combine these with techniques and ideas developed in the fields of genetics, molecular biology and biophysics. There has never been a hard-line between these disciplines in terms of content and technique. Today, the terms *molecular biology* and *biochemistry* are nearly interchangeable. *Biochemistry* is the study of the chemical substances and vital processes occurring in living organisms. Biochemists focus heavily on the role, function, and structure of biomolecules. The study of the chemistry behind biological processes and the synthesis of biologically active molecules are examples of biochemistry. *Genetics* is the study of the effect of genetic differences on organisms. Often this can be inferred by the absence of a normal component. The study of "mutants" – organisms with a changed gene that leads to the organism being different with respect to the so-called "wild type" or normal phenotype. Genetic interactions can often confound simple interpretations of such "knock-out" or "knock-in" studies. Molecular *biology* is the study of molecular underpinnings of the process of replication, transcription and translation of the genetic material. The central dogma of molecular biology where genetic material is transcribed into RNA and then translated into protein, despite being an oversimplified picture of molecular biology, still provides a good starting point for understanding the field. This picture, however, is undergoing revision in light of emerging novel roles for RNA. *Chemical Biology* seeks to develop new tools based on small molecules that allow minimal perturbation of biological systems while providing detailed information about their function. Further, chemical biology employs biological systems to create non-natural hybrids between biomolecules and synthetic devices.

This book caters to the needs for biochemistry students for a text which guides them through the rigmarole of molecules and their biochemical importance. Elucidating upon the basic concepts of both theory and application of molecular biochemistry, it seeks to understand the contemporary relevance of the field, particularly in the arena of genetic engineering, which emerged as one of the most exciting biological field in the 20th century and present century.

—Editor

1

Technological Applications

Gene Knockdown

The RNA interference pathway is often exploited in experimental biology to study the function of genes in cell culture and in vivo in model organisms. Double-stranded RNA is synthesized with a sequence complementary to a gene of interest and introduced into a cell or organism, where it is recognized as exogenous genetic material and activates the RNAi pathway. Using this mechanism, researchers can cause a drastic decrease in the expression of a targeted gene. Studying the effects of this decrease can show the physiological role of the gene product. Since RNAi may not totally abolish expression of the gene, this technique is sometimes referred as a "knockdown", to distinguish it from "knockout" procedures in which expression of a gene is entirely eliminated.

Extensive efforts in computational biology have been directed toward the design of successful dsRNA reagents that maximize gene knockdown but minimize "off-target" effects. Off-target effects arise when an introduced RNA has a base sequence that can pair with and thus reduce the expression of multiple genes at a time. Such problems occur more frequently when the dsRNA contains repetitive sequences. It has been estimated from studying the genomes of H. sapiens, C. elegans, and S. pombe that about 10% of possible siRNAs will have substantial off-target effects. A multitude of software tools have been developed implementing algorithms for the design of general, mammal-specific, and virus-specific siRNAs that are automatically checked for possible cross-reactivity.

Depending on the organism and experimental system, the exogenous RNA may be a long strand designed to be cleaved by dicer, or short RNAs designed to serve as siRNA substrates. In most mammalian cells, shorter RNAs are used because long double-stranded RNA molecules induce the mammalian interferon response, a form of innate immunity that reacts nonspecifically to foreign genetic material. Mouse oocytes and cells from early mouse embryos lack this reaction to exogenous dsRNA and are therefore a common model system for studying gene-knockdown effects in mammals. Specialized laboratory techniques have also been developed to improve the utility of RNAi in mammalian systems by avoiding the direct introduction of siRNA, for example, by stable transfection with a plasmid encoding the appropriate sequence from which siRNAs can be transcribed, or by more elaborate lentiviral vector systems allowing the inducible activation or deactivation of transcription, known as conditional RNAi.

Functional Genomics

Most functional genomics applications of RNAi in animals have used C. elegans and Drosophila, as these are the common model organisms in which RNAi is most effective. C. elegans is particularly useful for RNAi research for two reasons: firstly, the effects of the gene silencing are generally heritable, and secondly because delivery of the dsRNA is extremely simple. Through a mechanism whose details are poorly understood, bacteria such as E. coli that carry the desired dsRNA can be fed to the worms and will transfer their RNA payload to the worm via the intestinal tract. This "delivery by feeding" is just as effective at inducing gene silencing as more costly and time-consuming delivery methods, such as soaking the worms in dsRNA solution and injecting dsRNA into the gonads. Although delivery is more difficult in most other organisms, efforts are also underway to undertake large-scale genomic screening applications in cell culture with mammalian cells.

Figure: *A normal adult Drosophila fly, a common model organism used in RNAi experiments.*

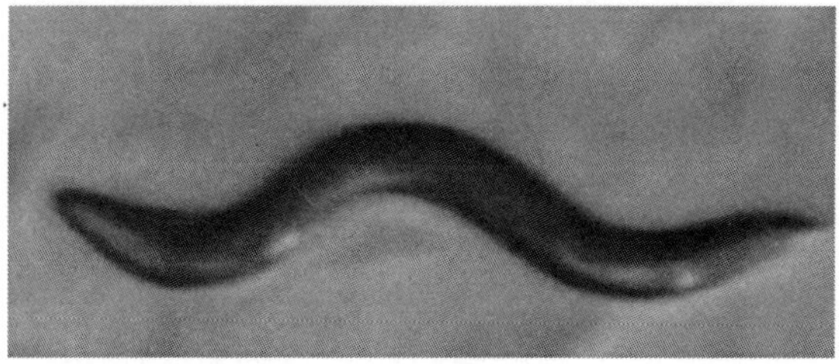

Figure: *An adult C. elegans worm, grown under RNAi suppression of a nuclear hormone receptor involved in desaturase regulation. These worms have abnormal fatty acid metabolism but are viable and fertile.*

Approaches to the design of genome-wide RNAi libraries can require more sophistication than the design of a single siRNA for a defined set of experimental conditions. Artificial neural networks are frequently used to design siRNA libraries and to predict their likely efficiency at gene knockdown. Mass genomic screening is widely seen as a promising method for genome annotation and has triggered the development of high-throughput screening methods based on microarrays. However, the utility of these screens and the ability of techniques developed on model organisms to generalize to even closely related species has been questioned, for example from C. elegans to related parasitic nematodes.

Functional genomics using RNAi is a particularly attractive technique for genomic mapping and annotation in plants because many plants are polyploid, which presents substantial challenges for more traditional genetic engineering methods. For example, RNAi has been successfully used for functional genomics studies in bread wheat (which is hexaploid) as well as more common plant model systems Arabidopsis and maize.

Medicine

It may be possible to exploit RNA interference in therapy. Although it is difficult to introduce long dsRNA strands into mammalian cells due to the interferon response, the use of short interfering RNA mimics has been more successful. Among the first applications to reach clinical trials were in the treatment of macular degeneration and respiratory syncytial virus, RNAi has also been shown to be effective in the reversal of induced liver failure in mouse models.

Other proposed clinical uses center on antiviral therapies, including topical microbicide treatments that use RNAi to treat infection (at Harvard University Medical School; in mice, so far) by herpes simplex virus type 2 and the inhibition of viral gene expression in cancerous cells, knockdown of host receptors and coreceptors for HIV, the silencing of hepatitis A and hepatitis B genes, silencing of influenza gene expression, and inhibition of measles viral replication. Potential treatments for neurodegenerative diseases have also been proposed, with particular attention being paid to the polyglutamine diseases such as Huntington's disease. RNA interference is also often seen as a promising way to treat cancer by silencing genes differentially upregulated in tumor cells or genes involved in cell division. A key area of research in the use of RNAi for clinical applications is the development of a safe delivery method, which to date has involved mainly viral vector systems similar to those suggested for gene therapy.

Despite the proliferation of promising cell culture studies for RNAi-based drugs, some concern has been raised regarding the safety of RNA interference, especially the potential for "off-target" effects in which a gene with a coincidentally similar sequence to the targeted gene is also repressed. A computational genomics study estimated that the error rate of off-target interactions is about 10%. One major study of liver disease in mice led to high death rates in the experimental animals, suggested by researchers to be the result of "oversaturation" of the dsRNA pathway, due to the use of shRNAs that have to be processed in the nucleus and exported to the cytoplasm using an active mechanism.

All these are considerations that are under active investigation, to reduce their impact in the potential therapeutic applications for RNAi. RNA interference-based applications are being developed to target persistent HIV-1 infection. Viruses like HIV-1 are particularly difficult targets for RNAi-attack because they are escape-prone, which requires combinatorial RNAi strategies to prevent viral escape. The future of antiviral RNAi therapeutics is very promising, but it remains of critical importance to include many controls in pre-clinical test models to unequivocally demonstrate sequence-specific action of the RNAi inducers.

Biotechnology

RNA interference has been used for applications in biotechnology, particularly in the engineering of food plants that produce lower levels of natural plant toxins. Such techniques take advantage of the stable and heritable RNAi phenotype in plant stocks. For example, cotton seeds are rich in dietary protein but naturally contain the toxic terpenoid product gossypol, making them unsuitable for human consumption. RNAi has been used to produce cotton stocks whose seeds contain reduced levels of delta-cadinene synthase, a key enzyme in gossypol production, without affecting the enzyme's production in other parts of the plant, where gossypol is important in preventing damage from plant pests. Similar efforts have been directed toward the reduction of the cyanogenic natural product linamarin in cassava plants.

Although no plant products that use RNAi-based genetic engineering have yet passed the experimental stage, development efforts have successfully reduced the levels of allergens in tomato plants and decreased the precursors of likely carcinogens in tobacco plants. Other plant traits that have been engineered in the laboratory include the production of non-narcotic natural products by the opium poppy, resistance to common plant viruses, and fortification of plants such as tomatoes with dietary antioxidants. Previous commercial products, including the Flavr Savr tomato and two cultivars of ringspot-resistant papaya, were originally developed using antisense technology but likely exploited the RNAi pathway.

Genome-scale RNAi Screening

The genome-scale RNAi research relies on the high-throughput screening (HTS) technology. The RNAi HTS technology allows genome-wide loss-of-function screening and is broadly used in the identification of genes associated with specific biological phenotypes. This technology has been hailed as the second genomics wave, following the first genomics wave of gene expression microarray and single nucleotide polymorphism discovery platforms. One of the major advantages of the genome-scale RNAi screening is its ability to simultaneously interrogate thousands of genes. With the ability of generating a large amount of data per experiment, the genome-scale RNAi screening has led to an explosion in the rate of generating data. Consequently, one of the most fundamental challenges in the genome-scale RNAi researches is to glean biological significance from mounds of data, which requires the adoption of suitable statistics/bioinformatics methods such as those presented in a recently published book for analyzing RNAi screens. The basic process of cell-based RNAi screening includes (i) the choice of an RNAi library, (ii) the selection of a robust and stable type of cells, (iii) the transfection of the selected cells with RNAi agents from the chosen RNAi library, (iv) necessary treatment or incubation, (v) signal detection, (vi) statistical and bioinformatics analysis, and (vii) the determination of important genes or therapeutical targets.

History and Discovery

Example petunia plants in which genes for pigmentation are silenced by RNAi. The left plant is wild-type; the right plants contain transgenes that induce suppression of both transgene and endogenous gene expression, giving rise to the unpigmented white areas of the flower.

The discovery of RNAi was preceded first by observations of transcriptional inhibition by antisense RNA expressed in transgenic plants, and more directly by reports of unexpected outcomes in experiments performed by plant scientists in the United States and the Netherlands in the early 1990s. In an attempt to alter flower colors in petunias, researchers introduced additional copies of a gene encoding chalcone synthase, a key enzyme for flower pigmentation into petunia plants of normally pink or violet flower color. The overexpressed gene was expected to result in darker flowers, but instead produced less pigmented, fully or partially white flowers, indicating that the activity of chalcone synthase had been substantially decreased; in fact, both the endogenous genes and the transgenes were downregulated in the white flowers. Soon after, a related event termed quelling was noted in the fungus Neurospora crassa, although it was not immediately recognized as related. Further investigation of the phenomenon in plants indicated that the downregulation was due to post-transcriptional inhibition of gene expression via an increased rate of mRNA degradation. This phenomenon was called co-suppression of gene expression, but the molecular mechanism remained unknown.

Not long after, plant virologists working on improving plant resistance to viral diseases observed a similar unexpected phenomenon. While it was known that plants expressing virus-specific proteins showed enhanced tolerance or resistance to viral infection, it was not expected that plants carrying only short, non-coding regions of viral RNA sequences would show similar levels of protection. Researchers believed that viral RNA produced by transgenes could also inhibit viral replication. The reverse experiment, in which short sequences of plant genes were introduced into viruses, showed that the targeted gene was suppressed in an infected plant. This phenomenon was labeled "virus-induced gene silencing" (VIGS), and the set of such phenomena were collectively called post transcriptional gene silencing.

After these initial observations in plants, many laboratories around the world searched for the occurrence of this phenomenon in other organisms. Craig C. Mello and Andrew Fire's 1998 Nature paper reported a potent gene silencing effect after injecting double stranded RNA into C. elegans. In investigating the regulation of muscle protein production, they observed that neither mRNA nor antisense RNA injections had an effect on protein production, but double-stranded RNA successfully silenced the targeted gene. As a result of this work, they coined the term RNAi. Fire and Mello's discovery was particularly notable because it represented the first identification of the causative agent for the phenomenon. Fire and Mello were awarded the Nobel Prize in Physiology or Medicine in 2006 for their work.

Gene Therapy for Color Blindness

Gene therapy for color blindness is an experimental gene therapy aiming to convert congenitally colorblind individuals to trichromats by introducing a photopigment gene that they lack. Though partial color blindness is considered only a mild disability and is controversial whether it is even a disorder, it is a condition that affects many people, particularly males. Complete color blindness, or achromatopsia, is very rare but more severe. While never demonstrated in humans, animal studies have shown that it is possible to confer color vision by injecting a gene of the missing photopigment using gene therapy. As of 2011 there is no medical entity offering this treatment, and no clinical trials available for volunteers.

Color Blindness

The retina of the human eye contains photoreceptive cells called cones that allow color vision. A normal trichromat individual possesses three different types of cones to distinguish different colors within the visible spectrum from 380 nm to 740 nm. The three types of cones are designated L, M, and S cones, and each type is sensitive to a certain range of wavelength of light depending on what photopigment it contains. More specifically, the L cone absorbs around 560 nm, the M cone absorbs near 530 nm, and the S cone absorbs near 420 nm. Contrary to popular belief, the peak absorption frequency for L, M, and S cones do not exactly correspond to red, green, and blue wavelength. Rather, the peak frequency for the L cone is orange, yellowish green in M cones, and blue-violet in S cones. These cones transduce the absorbed light into electrical information to be relayed to neurons in the retina such as retinal bipolar cells and retinal ganglion cells, before reaching the brain.

The signals from different cones are added or subtracted from each other to process the color of incoming light. For instance, the color red stimulate L cones more than M cones, whereas the color green stimulates the L and M cones more than the S cones. The colors are perceived in an opponent process, such that red and green are perceived in opposition, as are blue and yellow, black and white.

The gene loci coding for the photopigments: M-opsin and L-opsin are located in close proximity within the X chromosome and are highly polymorphic. Among the population, some have a deleted gene for the M photopigment in the X chromosome (such as in deuteranopia), whereas others have a mutated form of the gene (such as in deuteranomaly). Individuals who can express only two types of opsins in the cones are called dichromats. Because males have only one copy of the X chromosome, dichromatism is much more prevalent among men. With only two types of cones, dichromats are less capable of distinguishing between two colors. In the most common form of color blindness, deuteranopes have difficulty discriminating between red and green color. This is shown by their poor performance in Ishihara test. Although dichromatism poses little problem for daily life, dichromats may find some color-coded diagrams and maps difficult to read.

Less common forms of dichromacy include protoanopia (lack of L-cones), and tritanopia (lack of S-cones). If a person lacks two types of photopigments, they are considered monochromats. People lacking the three types of photopigments are said to have complete color blindness or achromatopsia. Color blindness can also result from damages to the visual cortex in the brain.

Theory

Experiments using a variety of mammals (including primates) demonstrated that it is possible to confer color vision to animals by introducing an opsin gene that the animal previously lacked. Using a replication-defective recombinant adeno-associated virus (rAAV) as a vector, the cDNA of the opsin gene found in the L or M cones can be delivered to some fraction of the cones within the retina via subretinal injection. Upon gaining the gene, the cone begins to express the new photopigment. The effect of therapy lasts until the cones die or the inserted DNA is lost within the cones.

While gene therapy for humans has been ongoing with some success, a gene therapy for humans to gain color vision has not been attempted to date. However, demonstrations using several mammals (including primates such as a squirrel monkey) suggest that the therapy should be feasible for humans as well. It is also theoretically possible for trichromats to be "upgraded" to tetrachromats by introducing new opsin genes.

Motivation

The goal of the gene therapy is to make some of the cones in the retina of a dichromat individual to express the missing photopigment. Although partial color blindness is considered to be a mild disability and even an advantage under certain circumstances (such as spotting camouflaged objects), it can pose challenges for certain occupations such as driving or piloting an aircraft. More generally, color codes in maps and figures may be difficult to read for individuals with color blindness.

Because only a single gene codes for a photopigment and the gene is only expressed in the retina, it is a relatively easy condition to treat using gene therapy compared to other genetic diseases. However, there remains the question of whether the therapy is worthwhile, for an individual to undergo an invasive subretinal injection to temporarily treat a condition that is more of an inconvenience than a disorder. However, complete color blindness, or achromatopsia, is very rare but more severe. Indeed, achromats cannot see any color, have a strong photophobia (blindness in full sun), and a reduced visual acuity (generally 20/200 after correction). Moreover, the research may have strong implications toward genetic therapy of other cone diseases. Other cone diseases such as Leber's congenital amaurosis, cone-rod dystrophy, and certain types of maculopathies may be treatable using the same techniques as the gene therapy used for color blindness.

Research so Far

There has been ongoing research for gene therapy to treat Leber's congenital amaurosis, a genetic disorder in photoreceptors that can lead to vision loss and blindness. These treatments use AAV vector and is delivered in much the same way as the gene therapy for color blindness.

Jacobs et al. published their research in the journal: Science in 2007, on their work introducing a human L-cone photopigment in mice. Since the mice possess only S cones and M cones, they are dichromats. The researchers replaced M-opsin with a cDNA of L-opsin in the X chromosome of some mice. By breeding these "knock-in" transgenic mice, they generated heterozygous females with both an M cone and an L cone. These mice had improved range of color vision and have gained trichromacy, as tested by electroretinogram and behavioral tests. However, this is more difficult to apply in the form of gene therapy.

In a paper published in the journal: Visual Neuroscience by Mauck et al., researchers used recombinant AAV vector to introduce the green fluorescent protein (GFP) gene in the cones of gerbils. The genetic insert was designed to only be expressed in S or M cones, and the study observed the expression of GFP in vivo over time. The study characterized the time of onset of expression, and also demonstrated that gene expression could stabilize if a sufficiently high dose of the viral vector is given.

Mancuso et al. published their research in the journal: Nature in 2009, on converting adult dichromat squirrel monkeys into trichromats using gene therapy. New world monkeys such as squirrel monkeys lack the L-opsin gene and are incapable of discriminating between certain shades of red and green. The researchers used recombinant AAV vector to deliver a human L-opsin gene into the monkey's retina. Cones that gained the missing genes began expressing the new photopigment.

The researchers raised two possibilities if the therapy worked—that the monkeys would either remain dichromatic with greater sensitivity for longer wavelength of light, or they would become trichromats. Electroretinogram recordings demonstrated that they are able to discriminate blue-green from red-violet, and have indeed gained trichromacy. The treated monkeys were also more successful when their color vision was tested with a modified

Ishihara test. In 2007, Alexander JJ et al. used gene therapy to restore some of the sight of mice with achromatopsia. The results were positive for 80% of the mice treated. Moreover, a paper by Komáromy et al., published in 2010, deals with gene therapy for a form of achromatopsia in dogs. Cone function and day vision have been restored for at least 33 months in two young dogs with achromatopsia. However, this therapy was less efficient for older dogs.

Theoretical Questions

According to research by David H. Hubel and Torsten Wiesel, suturing shut one eye of monkeys at an early age resulted in an irreversible loss of vision in that eye, even after the suture was removed. The study concluded that the neural circuitry for vision is wired during a "critical period" in childhood, after which the visual circuitry can no longer be rewired to process new sensory input. Contrary to this finding, Mancuso et al.'s success in conferring trichromacy to adult squirrel monkeys suggests that it is possible to adapt the preexisting circuit to allow greater acuity in color vision. The researchers concluded that integrating the stimulus from the new photopigment as an adult was not analogous to vision loss following visual deprivation.

It is yet unknown how the animals that gain a new photopigment are perceiving the new color. While the article by Mancuso et al. states that the monkey has indeed gained trichromacy and gained the ability to discriminate between red and green, they claim no knowledge of how the animal internally perceives the sensation.

While red/green color blindness among deuteranopes can be treated by introducing M-opsin genes, rarer forms of color blindness such as tritanopia can in principle be treated as well. For tritanopia, the S-opsin gene must be introduced instead of M-opsin gene.

Challenges

Despite the success in animals, there still remain challenges to conducting gene therapy on humans for treating color blindness.

Safety

How to deliver the viral vector into the retina is probably the main obstacle to making gene therapy a practical treatment for color blindness. Because the virus has to be injected directly by using a needle to penetrate the sclera of the eye, the treatment may be highly unpleasant and is a risk for eye infection. Without a way to deliver the virus noninvasively, the treatment is rather risky for the benefit gained.

As of now there is no knowledge of how frequently the gene needs to be injected to maintain trichromacy among congenitally colorblind individuals. At the time of publication, Mancuso et al. reports that the treated squirrel monkeys have maintained 2 years of color vision after the treatment. If repeat injections are needed, there is also the concern of the body developing an immune reaction to the virus. If a body develops sensitivity to the viral vector, the success of the therapy could be jeopardized and/or the body may respond unfavorably. An editorial by J. Bennett points to Mancuso et al.'s use of an "unspecified postinjection corticosteroid therapy".

Bennett suggests that the monkeys may have experienced inflammation due to the injection. However, the AAV virus that is commonly used for this study is non-pathogenic, and the body is less likely to develop an immune reaction. Needless to say, an extensive review of the safety of the treatment must precede any human trials. Obviously, the subject should first be evaluated to identify which photopigment they need to gain trichromacy. Also, while gene therapy may treat congenital color blindness (such as dichromacy), it is not intended to treat non-retinal forms of color blindness such as damage to the visual cortex of the brain.

Other Challenges

As a way to introduce new genetic information to change a person's phenotype, a gene therapy for color blindness is open to the same ethical questions and criticisms as gene therapy in general. Given the large number of people with color blindness, there is also the question of whether color blindness is a disorder or a phenotypic variant in the human population. In fact, dichromacy can even be advantageous under certain circumstances such as when identifying a camouflaged object from the background. Furthermore, even if gene therapy succeeds in converting incomplete colorblind individuals to trichromats, the degree of satisfaction among the subjects is unknown, though who watches black and white television by choice now? The following sentence reflects bias of those not afflicted with color vision deficiency- There probably is little increase in performance on day-to-day tasks, and it is uncertain how the quality of life will improve (or worsen) after the therapy.

Life Extension

Life extension science, also known as anti-aging medicine, experimental gerontology, and biomedical gerontology, is the study of slowing down or reversing the processes of aging to extend both the maximum and average lifespan. Some researchers in this area, and "life extensionists" or "longevists" (who wish to achieve longer lives for themselves), believe that future breakthroughs in tissue rejuvenation with stem cells, molecular repair, and organ replacement (such as with artificial organs or xenotransplantations) will eventually enable humans to have indefinite lifespans (agerasia) through complete rejuvenation to a healthy youthful condition. The sale of putative anti-aging products such as nutrition, physical fitness, skin care, hormone replacements, vitamins, supplements and herbs is a lucrative global industry, with the US market generating about $50 billion of revenue each year. Medical experts state that the use of such products has not been shown to affect the aging process, and many claims of anti-aging medicine advocates have been roundly criticized by medical experts, including the American Medical Association. Bioethicists question whether and how the human lifespan should be extended.

Average and Maximum Lifespans

During the process of aging, an organism accumulates damage to macromolecules, its cells, its tissues and its organs. This accumulated damage is the result of oxidation damage to the cell contents caused by free radicals. The maximum life span for humans is currently maximized at approximately 120 years, whereas the maximum lifespan of a mouse, commonly used as a model in research on aging, is about four years. Genetic differences between humans and mice that may account for these different aging rates include efficiency of DNA repair, types and quantities of antioxidant enzymes, and different rates of free radical production. The most important and challenging factor remains the telomere limitation of each individual species.

Average lifespan in a population is lowered by infant and child mortality, which are frequently linked to infectious diseases or nutrition problems. Later in life, vulnerability to accidents and age-related chronic disease such as cancer or cardiovascular disease play an increasing role in mortality. Extension of expected lifespan can often be achieved by access to improved medical care, vaccinations, good diet, exercise and avoidance of hazards such as smoking.

Maximum lifespan is determined by the rate of aging for a species inherent in its genes and by environmental factors. One widely recognized method of extending maximum lifespan in organisms such as nematodes is calorie restriction. Another technique used evolutionary pressure such as breeding from only older members.

Theoretically, extension of maximum lifespan could be achieved by reducing the rate of aging damage, by periodic replacement of damaged tissues, or by molecular repair or rejuvenation of deteriorated cells and tissues and the enhancement of telomerase enzyme activity. Future research will be geared towards telomere repair strategies.

Current Anti-aging Strategies and Issues

Diets and Supplements

Much life extension research focuses on nutrition—diets or supplements—as a means to extend lifespan, although few of these have been systematically tested for significant longevity effects. The many diets promoted by anti-aging advocates are often contradictory. Two diets with different approaches and some support from scientific research are the Paleolithic diet and Caloric restriction.

The free-radical theory of aging suggests that antioxidant supplements, such as Vitamin C, Vitamin E, Q_{10}, lipoic acid, carnosine, and N-acetylcysteine, might extend human life. However, combined evidence from several clinical trials suggest that â-Carotene supplements and high doses of Vitamin E increase mortality rates. Other substances proposed to extend lifespan include oxytocin, insulin, human chorionic gonadotropin (hCG), and erythropoietin (EPO). Resveratrol is a sirtuin stimulant that appears to extend lifespan in simple organisms such as nematodes and short-lived fish.

Some supplements, including the minerals selenium or zinc have been reported to extend the lifespan of rats and mice, though none has been proven to do so in humans, and significant toxic effects were observed. Metformin may also extend life span in mice.

Hormone Treatments

The anti-aging industry offers several hormone therapies. Some of these have been criticized for possible dangers to the patient and a lack of proven effect. For example, the American Medical Association has been critical of some anti-aging hormone therapies.

Even if some recent clinical studies have shown that low-dose GH treatment for adults with GH deficiency changes the body composition by increasing muscle mass, decreasing fat mass, increasing bone density and muscle strength, improves cardiovascular parameters (i.e. decrease of LDL cholesterol), and affects the quality of life without significant side effects.

The evidence for use of growth hormone as an anti-aging therapy is mixed and based on animal studies. An early study suggested that supplementation of mice with growth hormone increased average life expectancy. Additional animal experiments have suggested that growth hormone may generally act to shorten maximum lifespan; knockout mice lacking the receptor for growth hormone live especially long. Furthermore, mouse models lacking the insulin-like growth factor also live especially long and have low levels of growth hormone. Scientific controversy regarding anti-aging nutritional supplementation and medicine

Some critics dispute the portrayal of aging as a disease. For example, Leonard Hayflick, who determined that fibroblasts are limited to around 50 cell divisions, reasons that aging is an unavoidable consequence of entropy. Hayflick and fellow biogerontologists Jay Olshansky and Bruce Carnes have strongly criticized the anti-aging industry in response to what they see as unscrupulous profiteering from the sale of unproven anti-aging supplements. Ethics and politics of anti-aging nutritional supplementation and medicine

Politics relevant to the substances of life extension pertain mostly to communications and availability. In the United States, product claims on food and drug labels are strictly regulated. The First Amendment (freedom of speech) protects third-party publishers' rights to distribute fact, opinion and speculation on life extension practices. Manufacturers and suppliers also provide informational publications, but because they market the substances, they are subject to monitoring and enforcement by the Federal Trade Commission (FTC), which polices claims by marketers. What constitutes the difference between truthful and false claims is hotly debated and is a central controversy in this arena.

Proposed Strategies of Life Extension

Nanotechnology

Future advances in nanomedicine could give rise to life extension through the repair of many processes thought to be responsible for aging. Raymond Kurzweil, a futurist and transhumanist, believes that advanced medical nanorobotics could completely remedy the effects of aging by 2030.

Cloning and Body Part Replacement

Some life extensionists suggest that therapeutic cloning and stem cell research could one day provide a way to generate cells, body parts, or even entire bodies (generally referred to as reproductive cloning) that would be genetically identical to a prospective patient. Recently, the US Department of Defense initiated a program to research the possibility of growing human body parts on mice. Complex biological structures, such as mammalian joints and limbs, have not yet been replicated. Dog and primate brain transplantation experiments were conducted in the mid-20th century but failed due to rejection and the inability to restore nerve connections. As of 2006, the implantation of bio-engineered bladders grown from patients' own cells has proven to be a viable treatment for bladder disease. Proponents of body part replacement and cloning contend that the required biotechnologies are likely to appear earlier than other life-extension technologies.

The use of human stem cells, particularly embryonic stem cells, is controversial. Opponents' objections generally are based on interpretations of religious teachings or ethical considerations. Proponents of stem cell research point out that cells are routinely formed and destroyed in a variety of contexts. Use of stem cells taken from the umbilical cord or parts of the adult body may not provoke controversy.

The controversies over cloning are similar, except general public opinion in most countries stands in opposition to reproductive cloning. Some proponents of therapeutic cloning predict the production of whole bodies, lacking consciousness, for eventual brain transplantation.

Cryonics

For cryonicists (advocates of cryopreservation), storing the body at low temperatures after death may provide an "ambulance" into a future in which advanced medical technologies may allow resuscitation and repair. They speculate cryogenic temperatures will minimize changes in biological tissue for many years, giving the medical community ample time to cure all disease, rejuvenate the aged and repair any damage that is caused by the cryopreservation process.

Cryonicists do not believe that legal death is "real death" because stoppage of heartbeat and breathing—the usual medical criteria for legal death—occur before biological death of cells and tissues of the body. Even at room temperature, cells may take hours to die and days to decompose. Although neurological damage occurs within 4–6 minutes of cardiac arrest, the irreversible neurodegenerative processes do not manifest for hours. Cryonicists state that rapid cooling and cardio-pulmonary support applied immediately after certification of death can preserve cells and tissues for long-term preservation at cryogenic temperatures. People, particularly children, have survived up to an hour without heartbeat after submersion in ice water. In one case, full recovery was reported after 45 minutes underwater. To facilitate rapid preservation of cells and tissue, cryonics "standby teams" are available to wait by the bedside of patients who are to be cryopreserved to apply cooling and cardio-pulmonary support as soon as possible after declaration of death.

No mammal has been successfully cryopreserved and brought back to life, and resuscitation from cryonics is not possible with current science. Some scientists still support the idea based on their expectations of the capabilities of future science.

Strategies for Engineered Negligible Senescence (SENS)

Another proposed life extension technology would combine existing and predicted future biochemical and genetic techniques. One such theoretical strategy proposes a cure for cancer, stem cell treatments, addition of new enzymes to the human body and moving mitochondrial DNA to the cellular nucleus. There is no scientific evidence that supports this strategy, and has been called pseudoscientific because its proposed techniques are speculative. Strategies for Engineered Negligible Senescence (SENS) is the name Aubrey de Grey gives to his proposal to research regenerative medical procedures to periodically repair all the age-related damage in the human body, thereby maintaining a youthful state indefinitely. The term first appeared in print in de Grey's 1999 book The Mitochondrial Free Radical Theory of Aging, and was later prefaced with the term "strategies" in the article Time to Talk SENS: Critiquing the Immutability of Human Aging De Grey argues for a "goal-directed rather than curiosity-driven" approach to the science of aging, and to this purpose he tentatively identifies seven "damages of aging" and their potential methods of treatments.

SENS has received media attention but has been criticized by some scientists. While many biogerontologists (scientists who study aging) find it "worthy of discussion" and SENS conferences feature important research in the field, some contend that de Grey's programme is too speculative given current scientific research, referring to it as "fantasy rather than science".

Framework

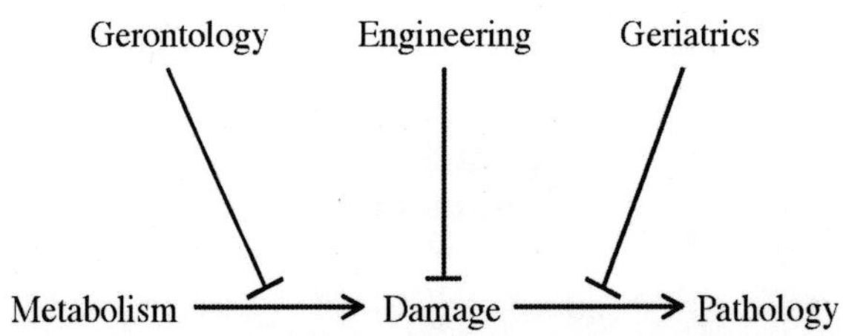

Figure: *The arrows with flat heads are a notation meaning "inhibits," used in the literature of gene expression and gene regulation.*

The goal of the Strategies for Engineered Negligible Senescence (SENS) is the complete reversal of all age-related illnesses and indefinite extension of the healthy human lifespan. The SENS project consists in implementing a series of periodic medical interventions designed to repair, prevent or render irrelevant all the types of molecular and cellular damage that cause age-related pathology and degeneration, in order to avoid debilitation and death from age-related causes.

De Grey defines aging as "the set of accumulated side effects from metabolism that eventually kills us", and, more specifically, as follows: "a collection of cumulative changes to the molecular and cellular structure of an adult organism, which result in essential metabolic processes, but which also, once they progress far enough, increasingly disrupt metabolism, resulting in pathology and death."

As de Grey states, "geriatrics is the attempt to stop damage from causing pathology; traditional gerontology is the attempt to stop metabolism from causing damage; and the SENS (engineering) approach is to eliminate the damage periodically, so keeping its abundance below the level that causes any pathology." His approach to biomedical gerontology ("anti-aging medicine") is thus distinctive because of its emphasis on rejuvenation rather than attempting to slow the aging process. He identifies what he believes to be the seven biological causes of senescence. According to de Grey, the causes of aging in humans are:

1. cell loss or atrophy (without replacement)
2. oncogenic nuclear mutations and epimutations,
3. cell senescence (Death-resistant cells),
4. mitochondrial mutations,
5. Intracellular junk or junk inside cells (lysosomal aggregates),
6. extracellular junk or junk outside cells (extracellular aggregates),
7. random extracellular cross-linking.

For each of these problems, de Grey outlines possible solutions, with a research and a clinical component. The clinical component is required because in some of the proposed therapies, feasibility has already been proven, but not completely applied and approved for use by human beings. He believes we will be able to apply these solutions before we completely understand the targeted aging mechanisms, which will take longer. He states that the goals work together to eliminate known causes of human senescence, are concrete, seem achievable, and are considered feasible by experts in the applicable fields. The goals were said to be taken from classical literature describing the biological causes of senescence.

De Grey proposes that engineered negligible senescence therapies could extend the human lifetime by many centuries or millennia, as early therapies give them enough time to seek more effective therapies later on. He describes an actuarial escape velocity of life extension, when advances in senescence treatment come rapidly enough to save the lives of the oldest beneficiaries of the previous treatments.

Types of Aging Damage and Treatment Schemes

Cancer-causing Nuclear Mutations/Epimutations—OncoSENS

These are changes to the nuclear DNA (nDNA), the molecule that contains our genetic information, or to proteins which bind to the nDNA. Certain mutations can lead to cancer, and, according to de Grey, non-cancerous mutations and epimutations do not contribute to aging within a normal lifespan, so cancer is the only endpoint of these types of damage that must be addressed. A mutation in a functional gene of a cell can cause that cell to malfunction or to produce a malfunctioning product, because of the sheer number of cells, de Grey believes that redundancy takes care of this problem, although cells that have mutated to produce toxic products might have to be disabled.

In de Grey's opinion, the effect of mutations and epimutations that really matters is cancer, this is because if even one cell turns into a cancer cell it might spread and become deadly. This would need to be corrected by a cure for cancer, if any is ever found. The SENS program focuses on a strategy called "whole-body interdiction of lengthening telomeres" (WILT), which would be made possible by periodic regenerative medicine treatments. Restoring telomeres would protect the ends of DNA from being cut off after successive divisions since each one normally removes some which is at first the telomere.

Mitochondrial Mutations—MitoSENS

Mitochondria are components in our cells that are important for energy production. They contain their own genetic material, and mutations to their DNA can affect a cell's ability to function properly. Indirectly, these mutations may accelerate many aspects of aging. Because of the highly oxidative environment in mitochondria and their lack of the sophisticated repair systems found in cell nucleus, mitochondrial mutations are believed to a be a major cause of progressive cellular degeneration. This would be corrected by allotopic expression—moving the DNA for mitochondria completely within the cellular nucleus, where it is better protected. In humans, all but 13 proteins are already protected in this way. De Grey argues that experimental evidence demonstrates that the operation is feasible. However, a 2003 study showed that some mitochondrial proteins are too hydrophobic to survive the transport from the cytoplasm to the mitochondria, and this is perhaps one of the reasons that not all of the mitochondrial genes have migrated to the nucleus during the course of evolution.

Intracellular Junk—LysoSENS

Our cells are constantly breaking down proteins and other molecules that are no longer useful or which can be harmful. Those molecules which can't be digested simply accumulate as junk inside our cells, which is readily detected in the form of lipofuscin granules. Atherosclerosis, macular degeneration, liver spots on the skin and all kinds of neurodegenerative diseases (such as Alzheimer's disease) are associated with this problem. Junk inside cells might be removed by adding new enzymes to the cell's natural digestion organ, the lysosome. These enzymes would be taken from bacteria, molds and other organisms that are known to completely digest animal bodies.

Extracellular Junk—AmyloSENS

Harmful junk protein can also accumulate outside of our cells. Junk outside cells might be removed by enhanced phagocytosis (the normal process used by the immune system), and small drugs able to break chemical beta-bonds. The large junk in this class can be removed surgically. Junk here means useless things accumulated by a body, but which cannot be digested or removed by its processes, such as the amyloid plaques characteristic of Alzheimer's disease and other amyloidoses. The oft-mentioned "toxins" that are identified as causes of many diseases most likely fit under this class.

Cell Loss and Atrophy—RepleniSENS

Some of the cells in our bodies cannot be replaced, or can be only replaced very slowly—more slowly than they die. This decrease in cell number affects some of the most important tissues of the body. Muscle cells are lost in skeletal muscles and the heart, causing them to become frailer with age. Loss of neurons in the substantia nigra causes Parkinson's disease, while loss of immune cells impairs the immune system. Cell depletion can be partly corrected by therapies involving exercise and growth factors. But stem cell therapy, regenerative medicine and tissue engineering are almost certainly required for any more than just partial replacement of lost cells. De Grey points out that this research of stem cell treatments is playing an increasingly important role in the international scientific community and progress is already occurring on many fronts. However, a large number of details are involved, and most such treatments are still experimental.

Cell Senescence—ApoptoSENS

This is a phenomenon where the cells are no longer able to divide, but also do not die and let others divide. They may also do other things that they are not supposed to do, like secreting proteins that could be harmful. Degeneration of joints, immune senescence, accumulation of visceral fat and type 2 diabetes are caused by this. Cells sometimes enter a state of resistance to signals sent, as part of a process called apoptosis, to instruct cells to destroy themselves. An example is the state known as cellular senescence. Cells in this state could be eliminated by forcing them to apoptose, and healthy cells would multiply to replace them. Cell killing with suicide genes or vaccines is suggested for making the cells undertake apoptosis.

Extracellular Crosslinks—GlycoSENS

Cells are held together by special linking proteins. When too many cross-links form between cells in a tissue, the tissue can lose its elasticity and cause problems including arteriosclerosis, presbyopia and weakened skin texture. These are chemical bonds between structures that are part of the body, but not within a cell. In senescent people many of these become brittle and weak. De Grey proposes to further develop small-molecular drugs and enzymes to break links caused by sugar-bonding, known as advanced glycation endproducts, and other common forms of chemical linking.

Scientific Controversy

While some fields mentioned as branches of SENS are broadly supported by the medical research community, i.e. stem cell research (RepleniSENS), anti-Alzheimers research (AmyloSENS) and oncogenomics (OncoSENS), the SENS programme as a whole has been a highly controversial proposal, with many critics arguing that the SENS agenda is fanciful and the highly complicated biomedical phenomena involved in the aging process contain too many unknowns for SENS to be scientific or implementable in the foreseeable future. Cancer may well deserve special attention as an aging-associated disease (OncoSENS), but the SENS claim that nuclear DNA damage only matters for aging because of cancer has been challenged.

In November 2005, 28 biogerontologists published a statement of criticism in EMBO reports, "Science fact and the SENS agenda: what can we reasonably expect from ageing research?," arguing "each one of the specific proposals that comprise the SENS agenda is, at our present stage of ignorance, exceptionally optimistic," and that some of the specific proposals "will take decades of hard work [to be medically integrated], if [they] ever prove to be useful."

The researchers argue that while there is "a rationale for thinking that we might eventually learn how to postpone human illnesses to an important degree," increased basic research, rather than the goal-directed approach of SENS, is presently the scientifically appropriate goal. This article was written in response to a July 2005 EMBO reports article previously published by de Grey and a response from de Grey was published in the same November issue. De Grey summarizes these events in "The biogerontology research community's evolving view of SENS," published on the Methuselah Foundation website.

De Grey Technology Review Controversy

In February 2005, Technology Review, which is owned by the Massachusetts Institute of Technology, published an article by Sherwin Nuland, a Clinical Professor of Surgery at Yale University and the author of "How We Die", that drew a skeptical portrait of Aubrey de Grey, at the time a computer associate in the Flybase Facility of the Department of Genetics at the University of Cambridge. While admiring de Grey's intelligence, Nuland concluded that he "would surely destroy us in attempting to preserve us" because living for such long periods would undermine what it means to be human. In the same issue of the magazine, Jason Pontin, the editor in chief and publisher of Technology Review, criticised de Grey. Writing an editor's letter, titled Against Transcendence, he questioned the usefulness and appropriateness of introducing transcendentalism, or transhumanism, into science. The April 2005 issue of Technology Review contained a reply by Aubrey de Grey and numerous comments from readers.

During June 2005 David Gobel, CEO and Co-founder of Methuselah Foundation offered Technology Review $20,000 to fund a prize competition to publicly clarify the viability of the SENS approach. In July 2005, Pontin announced a $20,000 prize, funded 50/50 by Methuselah Foundation and MIT Technology Review, open to any molecular biologist, with a record of publication in biogerontology, who could prove that SENS was "so wrong that it is unworthy of learned debate."

Technology Review received five submissions to its Challenge. In March, of 2006, Technology Review announced that it had chosen a panel of judges for the Challenge. Three met the terms of the prize competition. They were published by Technology Review on June 9, 2006. Accompanying the three submissions were rebuttals by de Grey, and counter-responses to de Grey's rebuttals. On July 11, 2006, Technology Review published the results of the SENS Challenge.

In the end, no one won the $20,000 prize. The judges felt that no submission met the criterion of the challenge and disproved SENS, although they unanimously agreed that one submission, by Preston Estep and his colleagues, was the most eloquent. Craig Venter succinctly expressed the prevailing opinion: "Estep et al.... have not demonstrated that SENS is unworthy of discussion, but the proponents of SENS have not made a compelling case for it." Summarizing the judges' deliberations, Pontin wrote, "SENS is highly speculative. Many of its proposals have not been reproduced, nor could they be reproduced with today's scientific knowledge and technology. Echoing Myhrvold, we might charitably say that de Grey's proposals exist in a kind of antechamber of science, where they wait (possibly in vain) for independent verification.

SENS does not compel the assent of many knowledgeable scientists; but neither is it demonstrably wrong." In a letter of dissent dated July 11, 2006 in Technology Review, Estep et al. criticized the ruling of the judges.

Social and Economic Implications

According to de Grey, of the roughly 150,000 people who die each day across the globe, about two thirds — 100,000 per day — die of age-related causes. In industrialized nations, the proportion is much higher, reaching 90%.

De Grey and other scientists in the general field have argued that the costs of a rapidly growing aging population will increase to the degree that the costs of an accelerated pace of aging research are easy to justify in terms of future costs avoided. Olshansky et al. 2006 argue, for example, that the total economic cost of Alzheimer's disease in the US alone will increase from $80–100 billion today to more than $1 trillion in 2050. "Consider what is likely to happen if we don't [invest further in aging research].

Take, for instance, the impact of just one age-related disorder, Alzheimer disease (AD). For no other reason than the inevitable shifting demographics, the number of Americans stricken with AD will rise from 4 million today to as many as 16 million by midcentury. This means that more people in the United States will have AD by 2050 than the entire current population of the Netherlands. Globally, AD prevalence is expected to rise to 45 million by 2050, with three of every four patients with AD living in a developing nation. The US economic toll is currently \$80–\$100 billion, but by 2050 more than \$1 trillion will be spent annually on AD and related dementias. The impact of this single disease will be catastrophic, and this is just one example."

Genetic Modification

Gene therapy, in which artificial genes are integrated with an organism to replace mutated or otherwise deficient genes, has been proposed as a future strategy to prevent aging. Targeting catalase to the mitochondria resulted in a 20% lifespan increase in transgenic mice, and improved performance in AAV therapeutically infected mice.

Fooling Genes

In The Selfish Gene, Richard Dawkins describes an approach to life-extension that involves "fooling genes" into thinking the body is young. Dawkins attributes inspiration for this idea to Peter Medawar. The basic idea is that our bodies are composed of genes that activate throughout our lifetimes, some when we are young, and others when we are older. Presumably, these genes are activated by environmental factors (this is known as epigenetics), and the changes caused by these genes activating can be lethal. It is a statistical certainty that we possess more lethal genes that activate in later life than in early life. Therefore, to extend life we should be able to prevent these genes from switching on, and we should be able to do so by "identifying changes in the internal chemical environment of a body that take place during aging... and by simulating the superficial chemical properties of a young body".

History of Life Extension and the Life Extension Movement

In 1970, the American Aging Association was formed under the impetus of Denham Harman, originator of the free radical theory of aging. Harman wanted an organization of biogerontologists that was devoted to research and to the sharing of information among scientists interested in extending human lifespan.

In 1976, futurists Joel Kurtzman and Philip Gordon wrote No More Dying. The Conquest Of Aging And The Extension Of Human Life, (ISBN 0-440-36247-4) the first popular book on research to extend human lifespan. Subsequently, Kurtzman was invited to testify before the House Select Committee on Aging, chaired by Claude Pepper of Florida, to discuss the impact of life extension on the Social Security system.

Saul Kent published The Life Extension Revolution (ISBN 0-688-03580-9) in 1980 and created a nutraceutical firm called the Life Extension Foundation, a non-profit organization that promotes dietary supplements. The Life Extension Foundation publishes a periodical called Life Extension Magazine. The 1982 bestselling book Life Extension: A Practical Scientific Approach (ISBN 0-446-51229-X) by Durk Pearson and Sandy Shaw further popularized the phrase "life extension".

In 1983, Roy Walford, a life-extensionist and gerontologist, published a popular book called Maximum Lifespan. In 1988, Walford and his student Richard Weindruch summarized their research into the ability of calorie restriction to extend the lifespan of rodents in The Retardation of Aging and Disease by Dietary Restriction (ISBN 0-398-05496-7). It had been known since the work of Clive McCay in the 1930s that calorie restriction can extend the maximum lifespan of rodents. But it was the work of Walford and Weindruch that gave detailed scientific grounding to that knowledge. Walford's personal interest in life extension motivated his scientific work and he practiced calorie restriction himself. Walford died at the age of 80 from complications caused by amyotrophic lateral sclerosis.

Money generated by the non-profit Life Extension Foundation allowed Saul Kent to finance the Alcor Life Extension Foundation, the world's largest cryonics organization. The cryonics movement had been launched in 1962 by Robert Ettinger's book, The Prospect of Immortality. In the 1960s, Saul Kent had been a co-founder of the Cryonics Society of New York. Alcor gained national prominence when baseball star Ted Williams was cryonically preserved by Alcor in 2002 and a family dispute arose as to whether Williams had really wanted to be cryopreserved.

Regulatory and legal struggles between the Food and Drug Administration (FDA) and the Life Extension Foundation included seizure of merchandise and court action. In 1991, Saul Kent and Bill Faloon, the principals of the Foundation, were jailed. The LEF accused the FDA of perpetrating a "Holocaust" and "seeking gestapo-like power" through its regulation of drugs and marketing claims.

In 1992, the American Academy of Anti-Aging Medicine (A4M) was formed to create what it considered an anti-aging medical specialty distinct from geriatrics, and to hold trade shows for physicians interested in anti-aging medicine. The American Board of Medical Specialties recognizes neither anti-aging medicine nor the A4M's professional standing.

Ethics and Politics of Life Extension

Leon Kass (chairman of the US President's Council on Bioethics from 2001 to 2005) has questioned whether potential exacerbation of overpopulation problems would make life extension unethical. He states his opposition to life extension with the words: "simply to covet a prolonged life span for ourselves is both a sign and a cause of our failure to open ourselves to procreation and to any higher purpose... [The] desire to prolong youthfulness is not only a childish desire to eat one's life and keep it; it is also an expression of a childish and narcissistic wish incompatible with devotion to posterity." John Harris, former editor-in-chief of the Journal of Medical Ethics, argues that as long as life is worth living, according to the person himself, we have a powerful moral imperative to save the life and thus to develop and offer life extension therapies to those who want them.

Transhumanist philosopher Nick Bostrom has argued that any technological advances in life extension must be equitably distributed and not restricted to a privileged few. In an extended metaphor entitled "The Fable of the Dragon-Tyrant", Bostrom envisions death as a monstrous dragon who demands human sacrifices. In the fable, after a lengthy debate between those who believe the dragon is a fact of life and those who believe the dragon can and should be destroyed, the dragon is finally killed. Bostrom argues that political inaction allowed many preventable human deaths to occur.

Aging as a Disease

Most mainstream medical organizations and practitioners do not consider aging to be a disease. David Sinclair says: "I don't see aging as a disease, but as a collection of quite predictable diseases caused by the deterioration of the body". The two main arguments used are that aging is both inevitable and universal while diseases are not. However not everyone agrees. Harry R. Moody, Director of Academic Affairs for AARP, notes that what is normal and what is disease strongly depends on a historical context. David Gems, Assistant Director of the Institute of Healthy Ageing, strongly argues that aging should be viewed as a disease. In response to the universality of aging, David Gems notes that it is as misleading as arguing that Basenji are not dogs because they do not bark. Because of the universality of aging he calls it a 'special sort of disease'. Robert M. Perlman, coined the terms 'aging syndrome' and 'disease complex' in 1954 to describe aging.

The discussion whether aging should be viewed as a disease or not has important implications. It would stimulate pharmaceutical companies to develop life extension therapies and it would also increase the regulation of the anti-aging market by the FDA. Anti-aging now falls under the regulations for cosmetic medicine which are less tight than those for drugs.

Pharmacological Gene Therapy

Pharmacological gene therapy is a new field that combines pharmacological therapy and gene therapy. It is used either to prevent a defective gene from producing its protein or to increase the concentration of normal protein produced in the body by insertion of DNA or RNA fragments. It can also be using to generate immunity from contagious disease, such as TB, via the process of DNA vaccination.

Gene Therapeutics can be used to treat such conditions as Cystic Fibrosis (via the addition of a normal CFTR gene), Haemophilia A or even some of the complications of AIDS. It uses a number of methods including: Gene augmentation Targeted Inhibition (using Antisense or antigene technology) Cell Killing (either direct or assisted) and DNA vaccination.

Predictive Medicine

Predictive medicine is a rapidly emerging field of medicine that entails predicting disease and instituting preventive measures in order to either prevent the disease altogether or significantly decrease its impact upon the patient (such as by preventing mortality or limiting morbidity). While different prediction methodologies exist, such as genomics, proteomics, and cytomics, the most fundamental way to predict future disease is based on genetics. Although proteomics and cytomics allow for the early detection of disease, much of the time those detect biological markers that exist because a disease process has already started. However, comprehensive genetic testing (such as through the use of DNA arrays or full genome sequencing) allows for the estimation of disease risk years to decades before any disease even exists, or even whether a healthy fetus is at higher risk for developing a disease in adolescence or adulthood. Individuals who are more susceptible to disease in the future can be offered lifestyle advice or medication with the aim of preventing the predicted illness. Current genetic testing guidelines supported by the health care professionals discourage purely predictive genetic testing of minors until they are competent to understand the relevancy of genetic screening so as to allow them to participate in the decision about whether or not it is appropriate for them. Genetic screening of newborns and children in the field of predictive medicine is deemed appropriate if there is a compelling clinical reason to do so, such as the availability of prevention or treatment as a child that would prevent future disease.

The Goal

The goal of predictive medicine is to predict future disease so that health care professionals and the patient themselves can be proactive in instituting lifestyle modifications and increased physician surveillance, such as bi-annual full body skin exams by a dermatologist or internist if their patient is found to have an increased risk of melanoma or an EKG and cardiology examination by a cardiologist if a patient is found to be at increased risk for a cardiac arrhythmia. Predictive medicine is intended not for patients but more so for healthy individuals, its purpose being to determine whether susceptibility to a particular disease is increased or not. A tremendous number of association studies have been published in scientific literature that show associations between specific genetic variants in a person's genetic code and a specific disease. Association and correlation studies have found that a female individual with a mutation in the BRCA1 gene has a 65% cumulative risk of breast cancer. Genetic variants in the Factor V gene is associated with an increased tendency to form blood clots, such as deep vein thrombosis (DVTs). Genetics tests are expected to reach the market more quickly than new medicines. Myriad Genetics is already generating revenue from genetic tests for BRCA1 and BRCA2.

Predictive medicine changes the fundamental paradigm of medicine from being reactive to being proactive and has the potential to significantly decrease the incidence and prevalence of both common and rare diseases.

Predictive Medicine

Available types of predictive medicine through health care professionals include:

- Newborn screening: Newborn screening is conducted just after birth to identify genetic disorders that can be treated early in life. This testing of infants for certain disorders is one of the most widespread uses of genetic screening - all US states currently test infants for phenylketonuria and congenital hypothyroidism. US state law mandates collecting a sample by pricking the heel of a newborn baby to obtain enough blood to fill a few circles on filter paper labeled with names of infant, parent, hospital, and primary physician.
- Diagnostic testing: Diagnostic testing is conducted to aid in the specificity diagnosis or detection of a disease. It is often used to confirm a particular diagnosis when a certain condition is suspected based on the subject's mutations and physical symptoms. The diversity in diagnostic testing ranges from common consulting room tests such as measuring blood pressure and urine tests to more invasive protocols such as biopsies.
- Medical bioinformatics: Medical bioinformatics involves determining individual cell molecular parameters by cytomics and single cell-based microarrays. A challenge remains to efficiently extract relevant predictive medicine parameters. This information is frequently extracted by computer-assisted identification and characterization of a few cell populations or gene clusters of interest.
- Prenatal testing: Prenatal testing is used to look for diseases and conditions in a fetus or embryo before it is born. This type of testing is offered for couples who have an increased risk of having a baby with

a genetic or chromosomal disorder. Screening can determine the sex of the fetus. Prenatal testing can help a couple decide whether to abort the pregnancy. Like diagnostic testing, prenatal testing can be noninvasive or invasive. Non-invasive techniques include examinations of the woman's womb through ultrasonography or maternal serum screens. These non-invasive techniques can evaluate risk of a condition, but cannot determine with certainty if the fetus has a condition. More invasive prenatal methods are slightly more risky for the fetus and involve needles or probes being inserted into the placenta or chorionic villus sampling.

- Carrier testing: Carrier testing is done to identify people who carry one copy of a gene mutation that, when present in both copies, causes a genetic disorder. This type of testing is offered to individuals who have genetic disorder in their family history or to people in ethnic groups with increased risk of certain genetic diseases. If both parents are tested, carrier testing can provide information about a couple's risk of having a child with a genetic disorder.
- Preconception testing: Preconception testing is conducted on a person or two potential parents before a child is even conceived. This allows prospective parents to understand the risk of diseases and likelihood of traits in their future offspring. Companies have already started to apply preconception testing through sperm banks and egg donation services. A company called Existence Genetics announced in 2010 that it has invented a new patent-pending preconception testing technology that they call the Pythia Approach (it's scientific name is Offspring Projections Through the Combined Analysis of Different Individuals, or OP-CADI) that uses a computer system to combine the genetic makeup of two potential parents in-order to predict the risk of a large number of common and rare diseases, as well as the likelihood of specific traits, in future offspring if those two potential parents were to have a child.

Health Benefits

The future of medicine's focus may potentially shift from treating existing diseases, typically late in their progression, to preventing disease before it sets in. Predictive medicine is based on probabilities: while it evaluates susceptibility to diseases, it is not able to predict with 100% certainty that a specific disease will occur. Unlike many preventive interventions that are directed at groups (e.g., immunization programs), predictive medicine is conducted on an individualized basis. For example, glaucoma is a monogenic disease whose early detection can allow to prevent permanent loss of vision. Predictive medicine is expected to be most effective when applied to polygenic multifactorial disease that are prevalent in industrialized countries, such as diabetes mellitus, hypertension, and myocardial infarction. With careful usage, predictive medicine methods such as genetic screens can help diagnose inherited genetic disease caused by problems with a single gene (such as cystic fibrosis) and help early treatment. Some forms of cancer and heart disease are inherited as single-gene diseases and some people in these high-risk families may also benefit from access to genetic tests. As more and more genes associated with increased susceptibility to certain diseases are reported, predictive medicine becomes more useful.

Direct-to-Consumer Genetic Testing

Direct-to-Consumer (DTC) genetic testing enables a consumer to screen his or her own genes without having to go though a health care professional. They can be ordered without the permission of a physician. Variety in DTC tests range from those testing for mutations associated with cystic fibrosis to breast cancer alleles. DTC tests make the applicability of predictive medicine very real and accessible to consumers. Benefits of DTC testing include this accessibility, privacy of genetic information, and promotion of proactive health care. Risks of obtaining DTC testing are the lack of governmental regulation and the interpreting of genetic information without professional counseling.

Limitations of Predictive Medicine

On a protein level, structure is more conserved than sequence. Therefore, in many diseases, having the faulty gene still does not necessarily mean someone will get the disease. Common, complex diseases in the wider population are affected not only by heredity, but also by external causes such as lifestyle and environment.

Therefore, genes are not perfect predictors of future health; individuals with both the high risk form of the gene and those without are all candidates to get the disease. Multiple factors in the environment, particular smoking, diet and exercise, infection, and pollution; play important roles and can be more important than genetic make-up.

This makes the results and risks determined by predictive medicine more difficult to quantify. Furthermore, the potential false positives or false negatives that may arise from a predictive genetic screen can cause substantial unnecessary strain on the individual.

Targeting medication to people who are genetically susceptible to a disease but do not yet show the symptoms of it can be a questionable measure. In large populations, there is concern that likely most of the people taking preventative medications would never have developed the disease anyway.

Many medications carry undesirable side effects that high risk individuals must then cope with. In contrast, several populations-based prevention measures (such as encouraging healthy diets or banning tobacco advertising) carry a far lower likelihood of adverse effects and are also less expensive.

Another potential downfall of commercially available genetic testing lies within the psychological impacts of accessibility to such data. For single-gene inherited diseases, counseling and the right to refuse a test (the right "not to know") have been found to be important.

However, adequate individual counseling can be difficult to employ to the potentially large proportion of the population likely to be identified as at high risk of common complex disease. Some people are vulnerable to adverse psychological reactions to genetic predictions of stigmatized or feared conditions, such as cancer or mental illness.

2

Synthesis by the Phosphoramidite Method

Building Blocks

Nucleoside phosphoramidites: As mentioned above, the naturally occurring nucleotides (nucleoside-3'- or 5'-phosphates) and their phosphodiester analogs are insufficiently reactive to afford an expedite synthetic preparation of oligonucleotides in high yields.

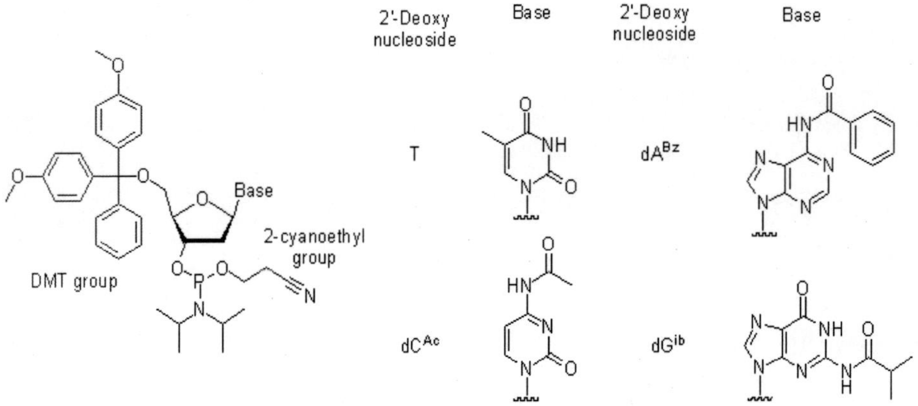

Figure: *Protected 2'-deoxynucleoside phosphoramidites.*

The selectivity and the rate of the formation of internucleosidic linkages is dramatically improved by using 3'-O-(N,N-diisopropyl phosphoramidite) derivatives of nucleosides (nucleoside phosphoramidites) that serve as building blocks in phosphite triester methodology. To prevent undesired side reactions, all other functional groups present in nucleosides have to be rendered unreactive (protected) by attaching protecting groups. Upon the completion of the oligonucleotide chain assembly, all the protecting groups are removed to yield the desired oligonucleotides. Below, the protecting groups currently used in commercially available and most common nucleoside phosphoramidite building blocks are briefly reviewed:

- The 5'-hydroxyl group is protected by an acid-labile DMT (4,4'-dimethoxytrityl) group.
- Thymine and uracil, nucleic bases of thymidine and uridine, respectively, do not have exocyclic amino groups and hence do not require any protection. Although the nucleic base of guanosine and 2'-deoxyguanosine does have an exocyclic amino group, its basicity is low to an extent that it does not react with phosphoramidites under the conditions of the coupling reaction. However, a phosphoramidite derived from the N2-unprotected 5'-O-DMT-2'-deoxyguanosine is poorly soluble in acetonitrile, the solvent commonly used in oligonucleotide synthesis. In contrast, the N2-protected versions of the same compound dissolve in acetonitrile well and hence are widely used. Nucleic bases adenine and cytosine bear the exocyclic amino groups reactive with the activated phosphoramidites under the conditions of the coupling reaction. Although, at the expense of additional steps in the synthetic cycle, the oligonucleotide chain assembly may be carried out using dA and dC phosphoramidites with unprotected amino groups, most often these are kept permanently protected over the entire length of the oligonucleotide chain assembly. The protection of the exocyclic amino groups have to be orthogonal to that of the 5'-hydroxy group because the latter is removed at the end of each synthetic cycle. The simplest to implement and hence the most widely accepted is the strategy where the exocyclic amino groups bear a base-labile protection. Most often, two protection schemes are used.
- In the first, the standard and more robust scheme, Bz (benzoyl) protection is used for A, dA, C, and dC, while G and dG are protected with isobutyryl group. More recently, Ac (acetyl) group is often used to protect C and dC.
- In the second, mild protection scheme, A and dA are protected with isobutyryl or phenoxyacetyl groups (PAC). C and dC bear acetyl protection, and G and dG are protected with 4-isopropylphenoxyacetyl (iPr-PAC) or dimethylformamidino (dmf) groups. Mild protecting groups are removed more readily than the standard protecting groups. However, the phosphoramidites bearing these groups are less stable when stored in solution.

- The phosphite group is protected by a base-labile 2-cyanoethyl group. Once a phosphoramidite has been coupled to the solid support-bound oligonucleotide and the phosphite moieties have been converted to the P(V) species, the presence of the phosphate protection is not mandatory for the successful conducting of further coupling reactions.

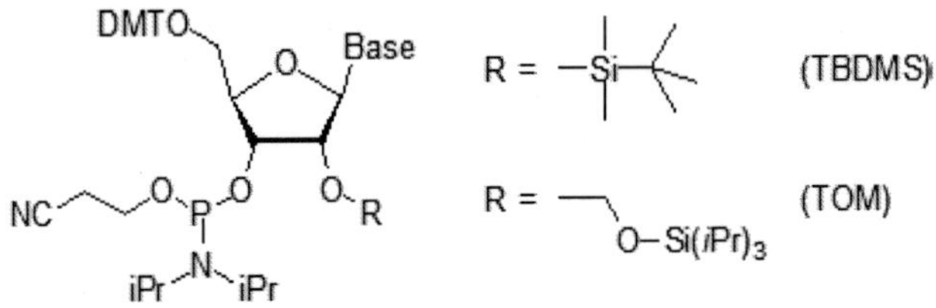

Figure: *2'-O-Protected ribonucleoside phosphoramidites.*

- In RNA synthesis, the 2'-hydroxy group is protected with TBDMS (t-butyldimethylsilyl) group. or with TOM (tri-iso-propylsilyloxymethyl) group, both being removable by treatment with fluoride ion.
- The phosphite moiety also bears a diisopropylamino (iPr_2N) group reactive under acidic conditions. Upon activation, the diisopropylamino group leaves to be substituted by the 5'-hydroxy group of the support-bound oligonucleotide.

Synthetic Cycle

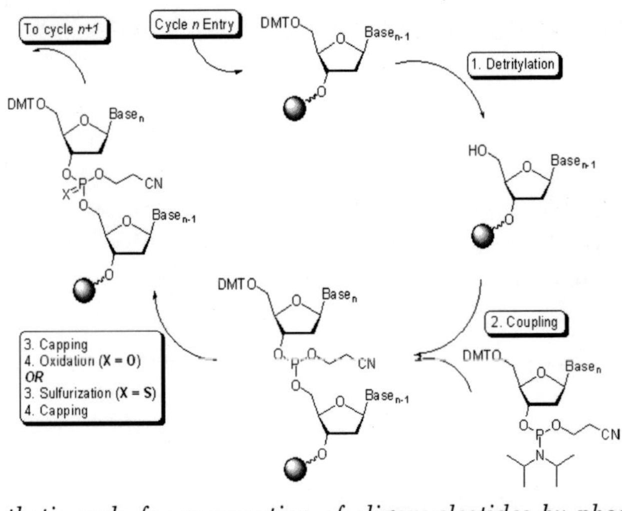

Figure: *Synthetic cycle for preparation of oligonucleotides by phosphoramidite method.*

Oligonucleotide synthesis is carried out by a stepwise addition of nucleotide residues to the 5'-terminus of the growing chain until the desired sequence is assembled. Each addition is referred to as a synthetic cycle (Scheme 6) and consists of four chemical reactions:

Step 1: De-blocking (detritylation): The DMT group is removed with a solution of an acid, such as 2% TCA or 3% Dichloroacetic acid (DCA), in an inert solvent (dichloromethane or toluene). The orange-colored DMT cation formed is washed out; the step results in the solid support-bound oligonucleotide precursor bearing a free 5'-terminal hydroxyl group. It is worth remembering that conducting detritylation for an extended time or with stronger than recommended solutions of acids leads to depurination of solid support-bound oligonucleotide and thus reduces the yield of the desired full-length product.

Step 2: Coupling: A 0.02 - 0.2M solution of nucleoside phosphoramidite (or a mixture of several phosphoramidites) in acetonitrile is activated by a 0.2 - 0.7M solution of an acidic azole catalyst, 1H-tetrazole, 2-ethylthiotetrazole, 2-benzylthiotetrazole, 4,5-dicyanoimidazole, or a number of similar compounds. The mixing is usually very brief and occurs in fluid lines of oligonucleotide synthesizers while the components are being delivered to the reactors containing solid support.

The activated phosphoramidite in 1.5 - 20-fold excess over the support-bound material is then brought in contact with the starting solid support (first coupling) or a support-bound oligonucleotide precursor (following couplings) whose 5'-hydroxy group reacts with the activated phosphoramidite moiety of the incoming nucleoside phosphoramidite to form a phosphite triester linkage. The phosphoramidite coupling is very rapid and requires, on small scale, about 20 s for its completion.

The reaction is also highly sensitive to the presence of water, particularly when dilute solutions of phosphoramidites are used, and is commonly carried out in anhydrous acetonitrile. Generally, the larger the scale of the synthesis, the lower the excess and the higher the concentration of the phosphoramidites is used. In contrast, the concentration of the activator is primarily determined by its solubility in acetonitrile and is irrespective of the scale of the synthesis. Upon the completion of the coupling, any unbound reagents and by-products are removed by washing.

Step 3: Capping: The capping step is performed by treating the solid support-bound material with a mixture of acetic anhydride and 1-methylimidazole or, less often, DMAP as catalysts and, in the phosphoramidite method, serves two purposes.

- After the completion of the coupling reaction, a small percentage of the solid support-bound 5'-OH groups (0.1 to 1%) remains unreacted and needs to be permanently blocked from further chain elongation to prevent the formation of oligonucleotides with an internal base deletion commonly referred to as (n-1) shortmers. The unreacted 5'-hydroxy groups are, to a large extent, acetylated by the capping mixture.

- It has also been reported that phosphoramidites activated with 1H-tetrazole react, to a small extent, with the O position of guanosine. Upon oxidation with I_2 /water, this side product, possibly via O -N7 migration, undergoes depurination. The apurinic sites thus formed are readily cleaved in the course of the final deprotection of the oligonucleotide under the basic conditions to give two shorter oligonucleotides thus reducing the yield of the full-length product. The O modifications are rapidly removed by treatment with the capping reagent as long as the capping step is performed prior to oxidation with I_2/water.
- The synthesis of oligonucleotide phosphorothioates does not involve the oxidation with I_2/water, and, respectively, does not suffer from the side reaction described above. On the other hand, if the capping step is performed prior to sulfurization, the solid support may contain the residual acetic anhydride and N-methylimidazole left after the capping step. The capping mixture interferes with the sulfur transfer reaction, which results in the extensive formation of the phosphate triester internucleosidic linkages in place of the desired PS triesters. Therefore, for the synthesis of OPS, it is advisable to conduct the sulfurization step prior to the capping step.

Step 4: Oxidation: The newly formed tricoordinated phosphite triester linkage is not natural and is of limited stability under the conditions of oligonucleotide synthesis.

The treatment of the support-bound material with iodine and water in the presence of a weak base (pyridine, lutidine, or collidine) oxidizes the phosphite triester into a tetracoordinated phosphate triester, a protected precursor of the naturally occurring phosphate diester internucleosidic linkage. This step is substituted with a sulfurization step to obtain oligonucleotide phosphorothioates. In the latter case, the sulfurization step is best carried out prior to capping.

Solid Supports

In solid-phase synthesis, an oligonucleotide being assembled is covalently bound, via its 3'-terminal hydroxy group, to a solid support material and remains attached to it over the entire course of the chain assembly.

The solid support is contained in columns whose dimensions depend on the scale of synthesis and may vary between 0.05 mL and several litres.

The overwhelming majority of oligonucleotides are synthesized on small scale ranging from 40 nmol to 1 ìmol. More recently, high-throughput oligonucleotide synthesis where the solid support is contained in the wells of multi-well plates (most often, 96 or 384 wells per plate) became a method of choice for parallel synthesis of oligonucleotides on small scale. At the end of the chain assembly, the oligonucleotide is released from the solid support and eluted from the column or the well.

Solid Support Material

In contrast to organic solid-phase synthesis and peptide synthesis, the synthesis of oligonucleotides proceeds best on non-swellable or low-swellable solid supports. The two most often used solid-phase materials are Controlled Pore Glass (CPG) and macroporous polystyrene (MPPS).

- CPG is commonly defined by its pore size. In oligonucleotide chemistry, pore sizes of 500, 1000, 1500, 2000, and 3000 Å are used to allow the preparation of about 50, 80, 100, 150, and 200-mer oligonucleotides, respectively. To make native CPG suitable for further processing, the surface of the material is treated with (3-aminopropyl)triethoxysilane to give Aminopropyl CPG. The aminopropyl arm may be further extended to result in Long Chain Aminoalkyl (LCAA) CPG. The amino group is then used as an anchoring point for linkers suitable for oligonucleotide synthesis.
- MPPS suitable for oligonucleotide synthesis is a low-swellable, highly cross-linked polystyrene obtained by polymerization of divinylbenzene (min 60%), styrene, and 4-chloromethylstyrene in the presence of a porogeneous agent. The macroporous chloromethyl MPPS obtained is converted to aminomethyl MPPS.

Linker Chemistry

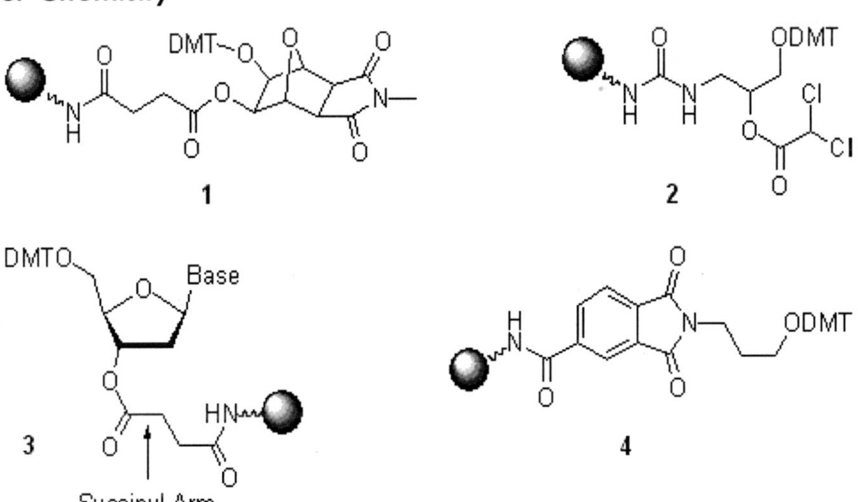

Figure: *Commercial solid supports for oligonucleotide synthesis.*

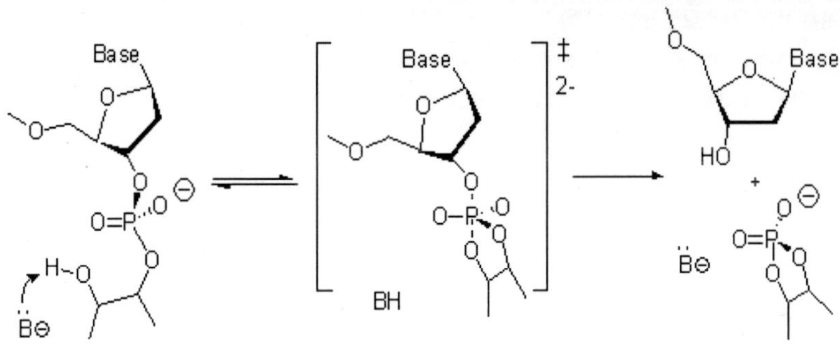

Figure: *Mechanism of 3'-dephosphorylation of oligonucleotides assembled on universal solid supports.*

To make the solid support material suitable for oligonucleotide synthesis, non-nucleosidic linkers or nucleoside succinates are covalently attached to the reactive amino groups in Aminopropyl CPG, LCAA CPG, or Aminomethyl MPPS.

The remaining unreacted amino groups are capped with acetic anhydride. Typically, three conceptually different groups of solid supports are used.

- Universal supports. In a more recent and more convenient method, the synthesis starts with the universal support where a non-nucleosidic linker is attached to the solid support material (Compounds 1 and 2). A phosphoramidite respective to the 3'-terminal nucleoside residue is coupled to the universal solid support in the first synthetic cycle of oligonucleotide chain assembly using the standard protocols. The chain assembly is then continued until the completion, after which the solid support-bound oligonucleotide is deprotected. The characteristic feature of the universal solid supports is that the release of the oligonucleotides occurs by the hydrolytic cleavage of a P-O bond that attaches the 3'-O of the 3'-terminal nucleotide residue to the universal linker as shown in Scheme 5. The critical advantage of this approach is that the same solid support is used irrespectively of the sequence of the oligonucleotide to be synthesized. For the complete removal of the linker and the 3'-terminal phosphate from the assembled oligonucleotide, the solid support 1 and several similar solid supports require gaseous ammonia, aqueous ammonium hydroxide, aqueous methylamine, or their mixture and are commercially available. The solid support 2 requires a solution of ammonia in anhydrous methanol and is also commercially available.
- Nucleosidic solid supports. In a historically first and still popular approach, the 3'-hydroxy group of the 3'-terminal nucleoside residue is attached to the solid support via, most often, 3'-O-succinyl arm as in compound 3. The oligonucleotide chain assembly starts with

the coupling of a phosphoramidite building block respective to the nucleotide residue second from the 3'-terminus. The 3'-terminal hydroxy group in oligonucleotides synthesized on nucleosidic solid supports is deprotected under the conditions somewhat milder than those applicable for universal solid supports. However, the fact that a nucleosidic solid support has to be selected in a sequence-specific manner reduces the throughput of the entire synthetic process and increases the likelihood of human error.

- Special solid supports are used for the attachment of desired functional or reporter groups at the 3'-terminus of synthetic oligonucleotides. For example, the commercial solid support 4 allows the preparation of oligonucleotides bearing 3'-terminal 3-aminopropyl linker. Similarly to non-nucleosidic phosphoramidites, many other special solid supports designed for the attachment of reactive functional groups, non-radioactive reporter groups, and terminal modifiers (e.c., hydrophobic tethers) and suited for various applications are commercially available.

Oligonucleotide Phosphorothioates and Their Synthesis

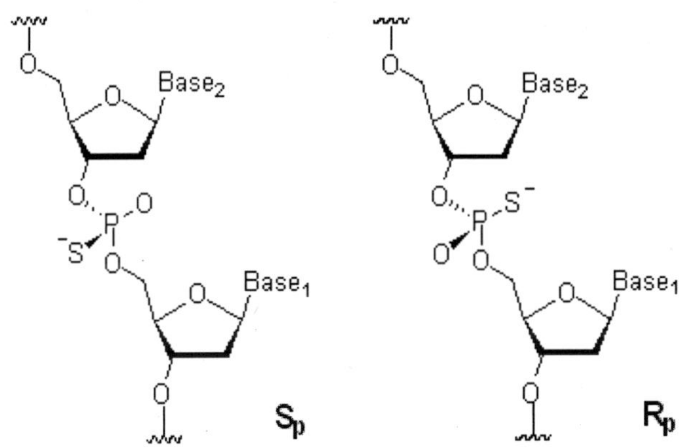

Figure: *S_p and R_p-diastereomeric internucleosidic phosphorothioate linkages.*

Oligonucleotide phosphorothioates (OPS) are modified oligonucleotides where one of the oxygen atoms in the phosphate moiety is replaced by sulfur. Only the phosphorothioates having sulfur at a non-bridging position as shown in figure are widely used and are available commercially. The replacement of the non-bridging oxygen with sulfur creates a new center of chirality at phosphorus. In a simple case of a dinucleotide, this results in the formation of a diastereomeric pair of S_p- and R_p-dinucleoside monophosphorothioates whose structures. In a n-mer oligonucleotide where all (n – 1) internucleosidic linkages are phosphorothioate linkages, the number of diastereomers m is calculated as m = 2· Being non-natural analogs of nucleic acids, OPS are substantially more stable towards hydrolysis by nucleases, the class of enzymes that destoy nucleic acids by breaking the bridging P-O bond of the phosphodiester moiety. This property determines the use of OPS as antisense oligonucleotides in in vitro and in vivo applications where the extensive exposure to nucleases is inevitable. Similarly, to improve the stability of siRNA, at least one phosphorothioate linkage is often introduced at the 3'-terminus of both sense and antisense strands. In chirally pure OPS, all-Sp diastereomers are more stable to enzymatic degradation than their all-Rp analogs. However, the preparation of chirally pure OPS remains a synthetic challenge. In laboratory practice, mixtures of diastereomers of OPS are commonly used.

Synthesis of OPS is very similar to that of natural oligonucleotides. The difference is that the oxidation step is replaced by sulfur transfer reaction (sulfurization) and that the capping step is performed after the sulfurization. Of many reported reagents capable of the efficient sulfur transfer, only three are commercially available:

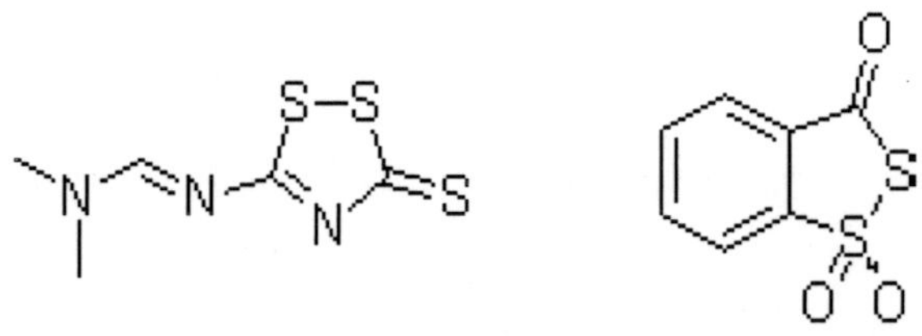

3 **4**

Figure: *Commercial sulfur transfer agents for oligonucleotide synthesis.*

- 3-((Dimethylaminomethylidene)amino)-3H-1,2,4-dithiazole-3-thione, DDTT (3, Fig. 4) provides rapid kinetics of sulfurization and high stability in solution. The reagent is available from several sources.
- 3H-1,2-benzodithiol-3-one 1,1-dioxide (4, Fig. 4) also known as Beaucage Reagent displays a better solubility in acetonitrile and short reaction times. However, the reagent is of limited stability in solution and is less efficient in sulfurizing RNA linkages.
- N,N,N'N'-Tetraethylthiuram disulfide (TETD) is soluble in acetonitrile and is commercially available. However, the sulfurization reaction of an internucleosidic DNA linkage with TETD requires 15 min, which is more than 10 times as slow as that with compounds 3 and 4.

Automation

In the past, oligonucleotide synthesis was carried out manually in solution or on solid phase. The solid phase synthesis was implemented using, as containers for the solid phase, miniature glass columns similar in their shape to low-pressure chromatography columns or syringes equipped with porous filters.

Currently, solid-phase oligonucleotide synthesis is carried out automatically using computer-controlled instruments (oligonucleotide synthesizers) and is technically implemented in column, multi-well plate, and array formats.

The column format is best suited for research and large scale applications where a high-throughput is not required. Multi-well plate format is designed specifically for high-throughput synthesis on small scale to satisfy the growing demand of industry and academia for synthetic oligonucleotides. A number of oligonucleotide synthesizers for small scale synthesis and medium to large scale synthesis are available commercially.

Synthesis of Oligonucleotide Microarrays

One may visualize an oligonucleotide microarray as a miniature multi-well plate where physical dividers between the wells (plastic walls) are intentionally removed. With respect to the chemistry, synthesis of oligonucleotide microarrays is different from the conventional oligonucleotide synthesis in two respects:

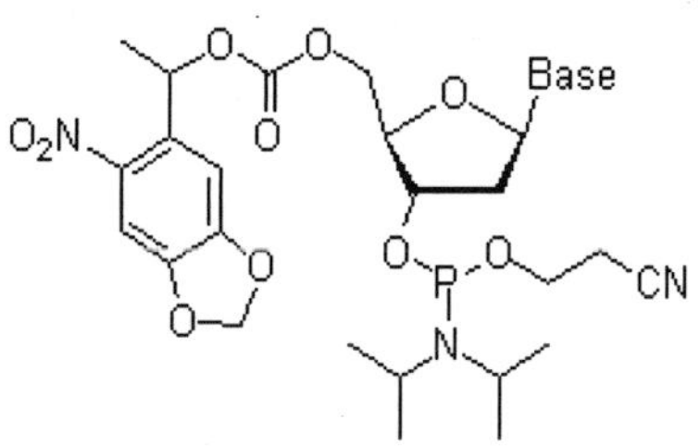

Figure: *5'-O-NPPOC-protected nucleoside phosphoramidite.*

- Oligonucleotides remain permanently attached to the solid phase, which requires the use of linkers that are stable under the conditions of the final deprotection procedure.
- The absence of physical dividers between the sites occupied by individual oligonucleotides, a very limited space on the surface of the microarray (one oligonucleotide sequence occupies a square 25x25 ìm) and the requirement of high fidelity of oligonucleotide synthesis dictate the use of site-selective 5'-deprotection techniques. In one approach, the removal of the 5'-O-DMT group is effected by electrochemical generation of the acid at the required site(s). In another approach, 5'-O-(á-methyl-6-nitropiperonyloxycarbonyl) (NPPOC) protecting group removable by UV-irradiation at 365 nm is used.

Post-synthetic Processing

After the completion of the chain assembly, the solid support-bound oligonucleotide is fully protected:

- The 5'-terminal 5'-hydroxy group is protected with DMT group;

- The internucleosidic phosphate or phosphorothioate moieties are protected with 2-cyanoethyl groups;
- The exocyclic amino groups in all nucleis bases except for T and U are protected with acyl protecting groups.

To furnish a functional oligonucleotide, all the protecting groups have to be removed. The N-acyl base protection and the 2-cyanoethyl phosphate protection may be, and is often removed simultaneously by treatment with inorganic bases or amines. However, the applicability of this method is limited by the fact that the cleavage of 2-cyanoethyl phosphate protection gives rise to acrylonitrile as a side product. Under the strong basic conditions required for the removal of N-acyl protection, acrylonitrile is capable of alkylation of nucleic bases, primarily, at the N3-position of thymine and uracil residues to give the respective N3-(2-cyanoethyl) adducts via Michael reaction. The formation of these side products may be avoided by treating the solid support-boud oligonucleotides with solutions of bases in an organic solvent, for instance, with 50% triethylamine in acetonitrile or 10% diethylamine in acetonitrile.

This treatment is strongly recommended for medium- and large scale preparations and is optional for syntheses on small scale where the concentration of acrylonitrile generated in the deprotection mixture is low.

Regardless of whether the phosphate protecting groups were removed first, the solid support-bound oligonucleotides are deprotected using one of the two general approaches.

- (1) Most often, 5'-DMT group is removed at the end of the oligonucleotide chain assembly. The oligonucleotides are then released from the solid phase and deprotected (base and phosphate) by treatment with aqueous ammonium hydroxide, aqueous methylamine, their mixtures, gaseous ammonia or methylamine or, less commonly, solutions of other primary amines or alkalies at ambient or elevated temperature. This removes all remaining protection groups from 2'-deoxyoligonucleotides, resulting in a reaction mixture containing the desired product. If the oligonucleotide contains any 2'-O-protected ribonucleotide residues, the deprotection protocol includes the second step where the 2'-O-protecting silyl groups are removed by treatment with fluoride ion by various methods.

The fully deprotected product is used as is, or the desired oligonucleotide can be purified by a number of methods. Most commonly, the crude product is desalted using ethanol precipitation, size exclusion chromatography, or reverse-phase HPLC. To eliminate unwanted truncation products, the oligonucleotides can be purified via polyacrylamide gel electrophoresis or anion-exchange HPLC followed by desalting.

- (2) The second approach is only used when the intended method of purification is reverse-phase HPLC. In this case, the 5'-terminal DMT group that serves as a hydrophobic handle for purification is kept on at the end of the synthesis. The oligonucleotide is deprotected under basic conditions as described above and, upon evaporation, is purified by reverse-phase HPLC. The collected material is then detritylated under aqueous acidic conditions. On small scale (less than 0.01–0.02 mmol), the treatment with 80% aqueous acetic acid for 15–30 min at room temperature is often used followed by evaporation of the reaction mixture to dryness in vacuo. Finally, the product is desalted as described above.
- For some applications, additional reporter groups may be attached to an oligonucleotide using a variety of post-synthetic procedures.

Characterization

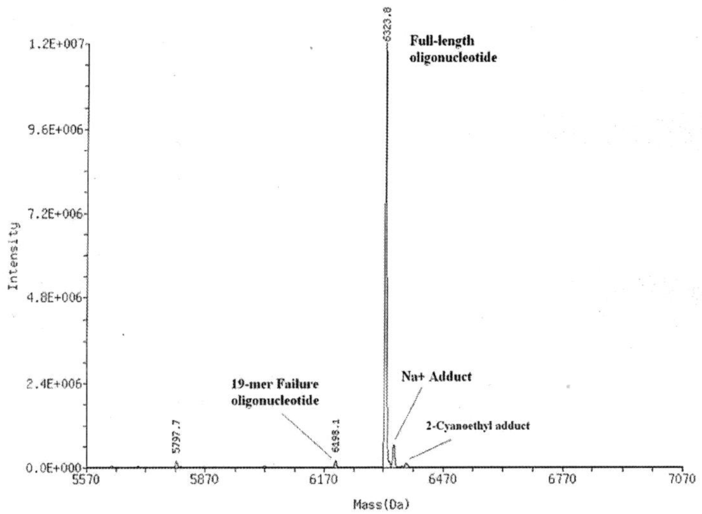

Figure: *Deconvoluted ES MS of crude oligonucleotide 5'-DMT-*T_{20} *(Calcd. 6324.26 Da).*

As with any other organic compound, it is prudent to characterize synthetic oligonucleotides upon their preparation. In more complex cases (research and large scale syntheses) oligonucleotides are characterized after their deprotection and after purification.

Although the ultimate approach to the characterization is sequencing, a relatively inexpensive and routine procedure, the considerations of the cost reduction preclude its use in routine manufacturing of oligonucleotides. In day-by-day practice, it is sufficient to obtain the molecular mass of an oligonucleotide by recording its mass spectrum.

Two methods are currently widely used for characterization of oligonucleotides: electrospray mass spectrometry (ES MS) and matrix-assisted laser desorption/ionization time-of-flight mass spectrometry (MALDI-TOF).

Prior to submitting a sample to the analysis by either of the methods, it is very important to exchange all metal ions that might be present in the sample for ammonium or trialkylammonium (e.c. triethylammonium) ions.

- In ES MS spectrum, a given oligonucleotide generates a set of ions that correspond to different ionization states of the compound. Thus, the oligonucleotide with molecular mass M generates ions with masses (M – nH)/n where M is the molecular mass of the oligonucleotide in the form of a free acid (all negative charges of internucleosidic phosphodiester groups are neutralized with H), n is the ionization state, and H is the atomic mass of hydrogen (1 Da). Most useful for characterization are the ions with n ranging from 2 to 5. Software supplied with the more recently manufactured instruments is capable of performing a deconvolution procedure that is, it finds peaks of ions that belong to the same set and derives the molecular mass of the oligonucleotide.
- To obtain more detailed information on the impurity profile of oligonucleotides, liquid chromatography-mass spectrometry (LC-MS or HPLC-MS) or capillary electrophoresis mass spectrometry (CEMS) are used.
- Antisense

In molecular biology and genetics, sense is a concept used to compare the polarity of nucleic acid molecules, such as DNA or RNA, to other nucleic acid molecules.

Depending on the context within molecular biology, sense may have slightly different meanings.

DNA Sense

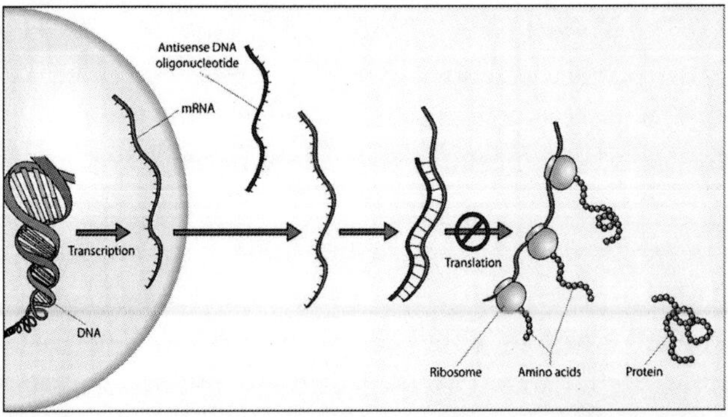

Figure: *Schematic showing how antisense DNA strands can interfere with protein translation.*

Molecular biologists call a single strand of DNA sense (or positive (+) sense) if an RNA version of the same sequence is translated or translatable into protein. Its complementary strand is called antisense (or negative (-) sense). Sometimes the phrase coding strand is encountered; however, protein coding and non-coding RNA's can be transcribed similarly from both strands, in some cases being transcribed in both directions from a common promoter region, or being transcribed from within introns, on both strands.

Antisense DNA

The two complementary strands of double-stranded DNA (dsDNA) are usually differentiated as the "sense" strand and the "antisense" strand. The DNA sense strand looks like the messenger RNA (mRNA) and can be used to read the expected protein code by human eyes (e.g. ATG codon = Methionine amino acid). However, the DNA sense strand itself is not used to make protein by the cell. It is the DNA antisense strand which serves as the source for the protein code, because, with bases complementary to the DNA sense strand, it is used as a template for the mRNA. Since transcription results in an RNA product complementary to the DNA template strand, the mRNA is complementary to the DNA antisense strand. The mRNA is what is used for translation (protein synthesis).

Hence, a base triplet 3'-TAC-5' in the DNA antisense strand can be used as a template which will result in an 5'-AUG-3' base triplet in mRNA (AUG is the codon for Methionine, the start codon). The DNA sense strand will have the triplet ATG which looks just like AUG but will not be used to make Methionine because it will not be used to make mRNA. The DNA sense strand is called a "sense" strand not because it will be used to make protein (it won't be), but because it has a sequence that looks like the protein codon sequence.

Students may find this confusing if they misunderstand the meaning of "sense" and if they misunderstand complementation. To make things more confusing, before the convention was set, very early textbooks disagreed on the DNA strands called "sense" and "antisense."

In biology and research, short antisense molecules can interact with complementary strands of nucleic acids, modifying expression of genes.

Example with Double-stranded DNA

DNA strand 1: antisense strand (copied to)'! RNA strand (sense)

DNA strand 2: sense strand

Some regions within a double strand of DNA code for genes, which are usually instructions specifying the order of amino acids in a protein along with regulatory sequences, splicing sites, noncoding introns, and other complicating details. For a cell to use this information, one strand of the DNA serves as a template for the synthesis of a complementary strand of RNA. The template DNA strand is called the transcribed strand with antisense sequence and the mRNA transcript is said to be sense sequence (the complement of antisense). Because the DNA is double-stranded, the strand complementary to the antisense sequence is called non-transcribed strand and has the same sense sequence as the mRNA transcript (though T bases in DNA are substituted with U bases in RNA).

5'CGCTATAG CGTTTCAT 3'	DNA antisense strand (template/noncoding), Watson strand	Used as a template for transcription.
3'GCGATAT CGCAAAGTA 5'	DNA sense strand (nontemplate/coding), Crick strand	Complementary to the template strand.
3'GCGAUAUCG CAAAGUA 5'	mRNA Sense transcript	RNA strand that is transcribed from the noncoding (template/antisense) strand. Note : Except for the fact that all thymines are now uracils (T—>U), it is complementary to the noncoding (template/ antisense) DNA strand (identical to the coding (nontemplate/sense) DNA strand). Note There is an AUG start codon at the 5' end (although written backwards here).
5'CGCUAUAGC GUUUCAU 3'		mRNA Antisense transcript RNA strand that is transcribed from the coding (nontemplate/sense) strand. Note: Except for the fact that all thymines are now uracils (T—>U), it is complementary to the coding (nontemplate/sense) DNA strand (identical to the noncoding (template/ antisense) DNA strand.

A note on the confusion between "sense" and "antisense" strands: The strand names actually depend on which direction you are writing the sequence that contains the information for proteins (the "sense" information), not on which strand is on the top or bottom (that is arbitrary).

The only real biological information that is important for labeling strands is the location of the 5' phosphate group and the 3' hydroxyl group because these ends determine the direction of transcription and translation. A sequence 5' CGCTAT 3' is equivalent to a sequence written 3' TATCGC 5' as long as the 5' and 3' ends are noted.

If the ends are not labeled, convention is to assume that the sequence is written in the 5' to 3' direction. Good rule of thumb for figuring out the "sense" strand: Look for the start codon ATG (AUG in mRNA). In the table example, the sense mRNA has the AUG codon at the end (remember that translation proceeds in the 5' to 3' direction).

Ambisense

A single-stranded genome that contains both positive-sense and negative-sense is said to be ambisense. Bunya viruses have 3 single-stranded RNA (ssRNA) fragments containing both positive-sense and negative-sense sections; arenaviruses are also ssRNA viruses with an ambisense genome, as they have 2 fragments that are mainly negative-sense except for part of the 5' ends of the large and small segments of their genome.

Antisense RNA

Antisense RNA is an RNA transcript that is complementary to endogenous mRNA. In other words, it is a non-coding strand complementary to the coding sequence of RNA; this is similar to negative-sense viral RNA. Introducing a transgene coding for antisense RNA is a technique used to block expression of a gene of interest. Radioactively-labelled antisense RNA can be used to show the level of transcription of genes in various cell types. Some alternative antisense structural types are being experimentally applied as antisense therapy, with at least one antisense therapy approved for use in humans.

When mRNA forms a duplex with a complementary antisense RNA sequence, translation is blocked. This process is called RNA interference.

Antisense nucleic acid molecules have been used experimentally to bind to mRNA and prevent expression of specific genes. Antisense therapies are also in development; in the USA, the Food and Drug Administration (FDA) has approved a phosphorothioate antisense oligo, fomivirsen (Vitravene), for human therapeutic use.

Cells can produce antisense RNA molecules naturally, which interact with complementary mRNA molecules and inhibit their expression.

RNA Sense in Viruses

In virology, the genome of an RNA virus can be said to be either positive-sense, also known as a "plus-strand", or negative-sense, also known as a "minus-strand". In most cases, the terms sense and strand are used interchangeably, making such terms as positive-strand equivalent to positive-sense, and plus-strand equivalent to plus-sense. Whether a virus genome is positive-sense or negative-sense can be used as a basis for classifying viruses.

Positive-sense

Positive-sense (5' to 3') viral RNA signifies that a particular viral RNA sequence may be directly translated into the desired viral proteins. Therefore, in positive-sense RNA viruses, the viral RNA genome can be considered viral mRNA, and can be immediately translated by the host cell. Unlike negative-sense RNA, positive-sense RNA is of the same sense as mRNA.

Some viruses (e.g., Coronaviridae) have positive-sense genomes that can act as mRNA and be used directly to synthesize proteins without the help of a complementary RNA intermediate. Because of this, these viruses do not need to have an RNA polymerase packaged into the virion.

Negative-sense

Negative-sense (3' to 5') viral RNA is complementary to the viral mRNA and thus must be converted to positive-sense RNA by an RNA polymerase prior to translation. Negative-sense RNA (like DNA) has a nucleotide sequence complementary to the mRNA that it encodes. Like DNA, this RNA cannot be translated into protein directly. Instead, it must first be transcribed into a positive-sense RNA that acts as an mRNA. Some viruses (Influenza, for example) have negative-sense genomes and so must carry an RNA polymerase inside the virion.

Enzymes

Enzyme-dependent antisense includes forms dependent on RNase H activity to degrade target mRNA, including single-stranded DNA, RNA, and phosphorothioate antisense. The R1 plasmid hok/sok system is an example of mRNA antisense regulation process, through enzymatic degradation of the resulting RNA duplex. Double stranded RNA acts as enzyme-dependent antisense through the RNAi/siRNA pathway, involving target mRNA recognition through sense-antisense strand pairing followed by target mRNA degradation by the RNA-induced silencing complex (RISC).

Steric blocking antisense (RNase-H-independent antisense) interferes with gene expression or other mRNA-dependent cellular processes by binding to a target sequence of mRNA and getting in the way of other processes. Steric blocking antisense includes 2'-O alkyl (usually in chimeras with RNase-H dependent antisense), peptide nucleic acid (PNA), locked nucleic acid (LNA), and Morpholino antisense.

Antisense mRNA

Antisense RNA is a single-stranded RNA that is complementary to a messenger RNA (mRNA) strand transcribed within a cell. Antisense RNA may be introduced into a cell to inhibit translation of a complementary mRNA by base pairing to it and physically obstructing the translation machinery. This effect is therefore stoichiometric. An example of naturally occurring mRNA antisense mechanism is the hok/sok system of the E. coli R1 plasmid. Antisense RNA has long been thought of as a promising technique for disease therapy; the only such case to have reached the market is the drug fomivirsen. One commentator has characterized antisense RNA as one of "dozens of technologies that are gorgeous in concept, but exasperating in [commercialization]".

Generally, antisense RNA still lack effective design, biological activity, and efficient route of administration. Historically, the effects of antisense RNA have often been confused with the effects of RNA interference (RNAi), a related process in which double-stranded RNA fragments called small interfering RNAs trigger catalytically mediated gene silencing, most typically by targeting the RNA-induced silencing complex (RISC) to bind to and degrade the mRNA.

Attempts to genetically engineer transgenic plants to express antisense RNA instead activate the RNAi pathway, although the processes result in differing magnitudes of the same downstream effect, gene silencing. Well-known examples include the Flavr Savr tomato and two cultivars of ringspot-resistant papaya.

Transcription of longer cis-antisense transcripts is a common phenomenon in the mammalian transcriptome. Although the function of some cases have been described, such as the Zeb2/Sip1 antisense RNA, no general function has been elucidated. In the case of Zeb2/Sip1, the antisense noncoding RNA is opposite the 5' splice site of an intron in the 5'UTR of the Zeb2 mRNA. Expression of the antisense ncRNA prevents splicing of an intron that contains a ribosome entry site necessary for efficient expression of the Zeb2 protein. Transcription of long antisense ncRNAs is often concordant with the associated protein-coding gene, but more detailed studies have revealed that the relative expression patterns of the mRNA and antisense ncRNA are complex.

Cis-natural Antisense Transcript

Natural antisense transcripts (NATs) are a group of RNAs encoded within a cell that have transcript complementarity to other RNA transcripts. They have been identified in multiple eukaryotes, including humans, mice, yeast and Arabidopsis thaliana. This class of RNAs includes both protein-coding and non-coding RNAs. Current evidence has suggested a variety of regulatory roles for NATs, such as RNA interference (RNAi), alternative splicing, genomic imprinting, and X-chromosome inactivation.

NATs are broadly grouped into two categories based on whether they act in cis or in trans. Trans-NATs are transcribed from a different location than their targets and usually have complementarity to multiple transcripts with some mismatches. MicroRNAs (miRNA) are an example of trans-NATs that can target multiple transcripts with a few mismatches. Cis-natural antisense transcripts (cis-NATs) on the other hand are transcribed from the same genomic locus as their target but from the opposite DNA strand and form perfect pairs.

Orientation of cis-NATs

Cis-NATs have a variety of orientations and differing lengths of overlap between pairs. There have been five identified orientations for cis-NATs to date. The most common orientation is head-to-head, where the 5' ends of both transcripts align together. This orientation would result in the greatest knockdown of gene expression if transcriptional collision is the reason for transcript inhibition. There is however some studies that have suggested that tail-to-tail orientations are the most common NAT pairs. Others such as tail to tail, overlapping, nearby head-to- head, and nearby tail-to-tail are less frequently encountered. Completely overlapping NATs involve the antisense gene being located completely over top of each other. Nearby head-to-head and tail-to-tail orientations are physically discrete from each other but are located very close to each other. Current evidence suggests that there is an overrepresentation of NAT pairs in genes that have catalytic activity. There may be something about these genes in particular that makes them more prone to this type of regulation.

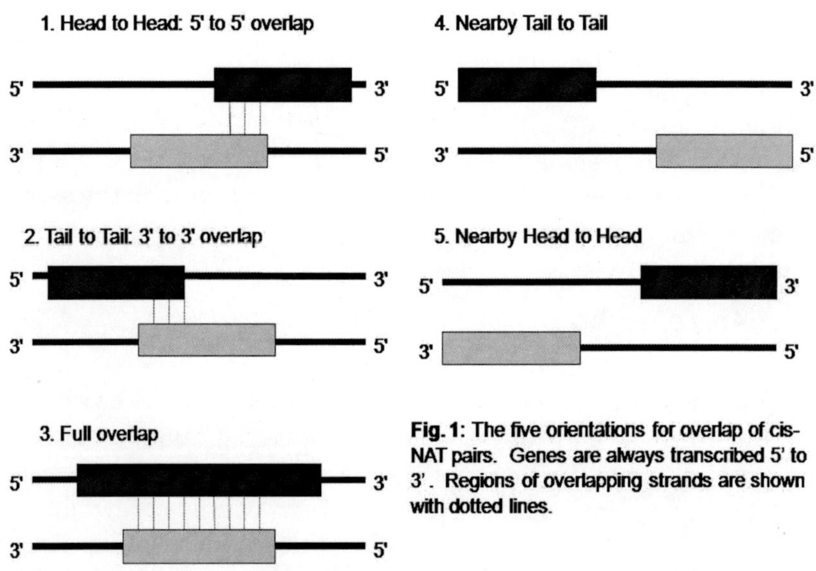

Fig. 1: The five orientations for overlap of cis-NAT pairs. Genes are always transcribed 5' to 3'. Regions of overlapping strands are shown with dotted lines.

Figure: *Orientations of cis-NATs within the genome*

Identification Approach

Identification of NATs in whole genomes is possible due to the large collection of sequence data available from multiple organisms. In silico methods for detecting NATs suffer from several shortcomings depending on the source of sequence information. Studies that use mRNA have sequences whose orientations are known, but the amount of mRNA sequence information available is small. Predicted gene models using algorithms trained to look for genes gives an increased coverage of the genome at the cost of confidence in the identified gene. Another resource is the extensive expressed sequence tag (EST) libraries but these small sequences must first be assigned an orientation before useful information can be extracted from them. Some studies have utilized special sequence information in the ESTs such as the poly(A) signal, poly(A) tail, and splicing sites to both filter the ESTs and to give them the correct transcriptional orientation. Combinations of the different sequence sources attempts to maximize coverage as well as maintain integrity in the data.

Pairs of NATs are identified when they form overlapping clusters. There is variability in the cut-off values used in different studies but generally ~20 nucleotides of sequence overlap is considered the minimum for transcripts to be considered and overlapping cluster. Also, transcripts must map to only one other mRNA molecule in order for it to be considered a NAT pair. Currently there are a variety of web and software resources that can be used to look for antisense pairs. The NATsdb or Natural Antisense Transcript database is a rich tool for searching for antisense pairs from multiple organisms.

Mechanisms

Molecular mechanisms behind the regulatory role of cis-NATs are not currently well understood. Three models have been proposed to explain the regulatory effects that cis-NATs have on gene expression. The first model attributes that base pairing between the cis-NAT and its complementary transcript result in a knockdown of mRNA expression. The assumption of this model is that there will be a precise alignment of at least 6 base pairs between the cis-NAT pair to make double stranded RNA. Epigenetic modifications like DNA methylation and post-translational modification of core histones form the basis of the second model.

Although it is not yet clearly understood, it is thought that the reverse transcript guides methylation complexes and/or histone-modifying complexes to the promoter regions of the sense transcript and cause an inhibition of expression from the gene. Currently it is not known what attributes of cis-NATs are crucial for the epigenetic model of regulation. The final proposed model that has gained favour due to recent experimental evidence is the transcriptional collision model. During the process of transcription of cis-NATs, the transcriptional complexes assemble in the promoter regions of the gene. RNA polymerases will then begin transcribing the gene at the transcription initiation site laying down nucleotides in a 5' to 3' direction. In the areas of overlap between the cis-NATs the RNA polymerases will collide and stop at the crash site. Transcription is inhibited because RNA polymerases prematurely stop and their incomplete transcripts get degraded.

Importance

Regulation of many biological processes such as development, metabolism and many others requires a careful co-ordination between many different genes; this is usually referred to as a gene regulatory network. A flurry of interest in gene regulatory networks has been sparked by the advent of sequenced genomes of multiple organisms. The next step is to use this information to figure out how genes work together and not just in isolation. During the processes of mammalian development, there is an inactivation of the extra X-chromosome in females. It has been shown that a NAT pair called Xist and Tsix are involved in the hypermethylation of the chromosome. As much as 20-30% of mammalian genes have been shown to be the targets of miRNAs, which highlights the importance of these molecules as regulators across a wide number of genes. Evolutionary reasons for utilizing RNA for regulation of genes may be that it is less costly and faster than synthesizing proteins not needed by the cell. This could have had a selective advantage for early eukaryotes with this type of transcriptional regulation.

Cis-NATs and Disease

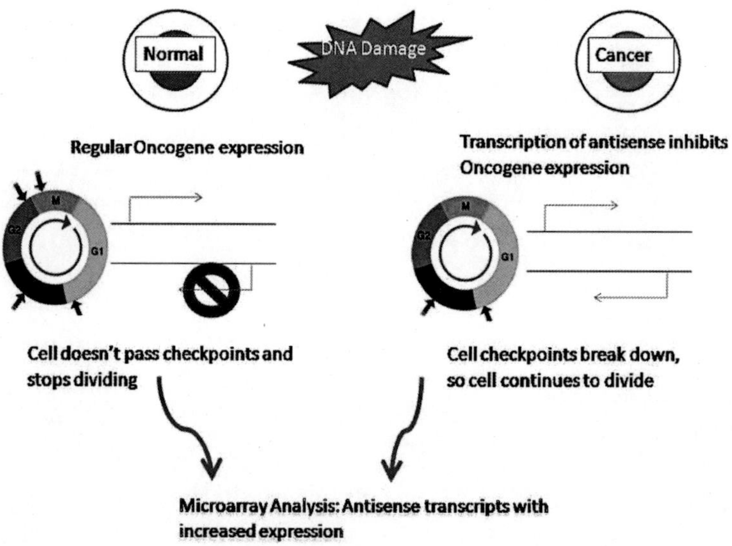

Figure: *Aberrant transcription of antisense transcripts can result in inhibition of oncogenes and allow cell to continue past cell cycle check points. Putative new oncogenes and tumor suppressor genes can be found by looking for upregulated antisense transcripts in cancer cells.*

Antisense transcription might contribute to disease through chromosomal changes that result in the production of aberrant antisense transcripts. A documented case of cis-NATs being involved in human disease comes from an inherited form of á-thalassemia where there is silencing of the hemoglobin á-2 gene through the action of a cis-NAT. It is thought that in malignant cancer cells with activated transposable elements creates a large amount of transcriptional noise. It is likely that aberrant antisense RNA transcripts resulting from this transcriptional noise may cause stochastic methylation of CpG islands associated with oncogenes and tumor suppressor genes. This inhibition would further progress the malignancy of the cells since they lose key regulator genes. By looking at upregulated antisense transcripts in tumor cells, researchers and look for more candidate tumor suppressor genes. Also, aberrant cis-NATs have been implicated in neurological diseases such as Parkinson's disease.

Morpholino

In molecular biology, a Morpholino is a molecule in a particular structural family that is used to modify gene expression. Morpholino oligomers (oligos) are an antisense technology used to block access of other molecules to specific sequences within nucleic acid. Morpholinos block small (~25 base) regions of the base-pairing surfaces of ribonucleic acid (RNA).

This article discusses only the Morpholino antisense oligos, which are nucleic acid analogs. The word "morpholino" can occur in other chemical names, referring to chemicals containing a six-membered morpholine ring. To help avoid confusion with other morpholine-containing molecules, when describing oligos "Morpholino" is often capitalized as a trade name, though this usage is not consistent throughout the scientific literature. Morpholinos are sometimes referred to as "PMO" (phosphorodiamidate morpholino oligo).

Morpholinos are usually used as a research tool for reverse genetics by knocking down gene function. This is achieved by preventing cells from making a targeted protein or by modifying the splicing of pre-mRNA. Knocking down gene expression is a powerful method for learning about the function of a particular protein; in similar manner, causing a specific exon to be spliced out of a protein can help to determine the function of the protein moiety encoded by that exon. These molecules have been applied to studies in several model organisms, including mice, zebrafish, frogs, and sea urchins.

Morpholinos are also in development as pharmaceutical therapeutics targeted against pathogenic organisms such as bacteria or viruses and for amelioration of genetic diseases. These synthetic oligos were conceived by James E. Summerton (Gene Tools, LLC) and developed in collaboration with Dwight D. Weller (AVI BioPharma Inc.).

Structure

Morpholinos are synthetic molecules that are the product of a redesign of natural nucleic acid structure. Usually 25 bases in length, they bind to complementary sequences of RNA by standard nucleic acid base-pairing. Structurally, the difference between Morpholinos and DNA is that while Morpholinos have standard nucleic acid bases, those bases are bound to morpholine rings instead of deoxyribose rings and linked through phosphorodiamidate groups instead of phosphates. This may be easiest to visualize by referring to the first figure and comparing the structures of the two strands depicted there, one of RNA and the other of a Morpholino. Replacement of anionic phosphates with the uncharged phosphorodiamidate groups eliminates ionization in the usual physiological pH range, so Morpholinos in organisms or cells are uncharged molecules. The entire backbone of a Morpholino is made from these modified subunits. Morpholinos are most commonly used as single-stranded oligos, though heteroduplexes of a Morpholino strand and a complementary DNA strand may be used in combination with cationic cytosolic delivery reagents.

Function

Morpholinos do not degrade their target RNA molecules, unlike many antisense structural types (e.g., phosphorothioates, siRNA). Instead, Morpholinos act by "steric blocking", binding to a target sequence within an RNA and simply getting in the way of molecules that might otherwise interact with the RNA. Morpholino oligos are often used to investigate the role of a specific mRNA transcript in an embryo. Developmental biologists inject Morpholino oligos into eggs or embryos of zebrafish, African clawed frog (Xenopus), sea urchin, and killifish (F. heteroclitus) producing morphant embryos, or electroporate Morpholinos into chick embryos at later stages of development. With appropriate cytosolic delivery systems, Morpholinos are effective in cell culture. Morpholinos are being developed as pharmaceuticals under the name "NeuGenes" by AVI BioPharma Inc. They have been used in mammals such as mice and some are currently being tested in clinical trials for treatment of Duchenne muscular dystrophy in humans.

Normal Gene Expression in Eukaryotes

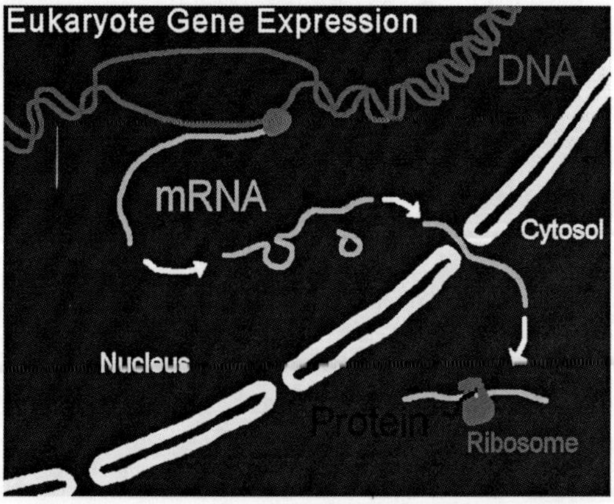

Figure: *Eukaryotic gene expression without intervention by a Morpholino*

In eukaryotic organisms, pre-mRNA is transcribed in the nucleus, introns are spliced out, then the mature mRNA is exported from the nucleus to the cytoplasm. The small subunit of the ribosome usually starts by binding to one end of the mRNA and is joined there by various other eukaryotic initiation factors, forming the initiation complex. The initiation complex scans along the mRNA strand until it reaches a start codon, and then the large subunit of the ribosome attaches to the small subunit and translation of a protein begins. This entire process is referred to as gene expression; it is the process by which the information in a gene, encoded as a sequence of bases in DNA, is converted into the structure of a protein. A Morpholino can modify splicing or block translation, depending on the Morpholino's base sequence.

Blocking Translation

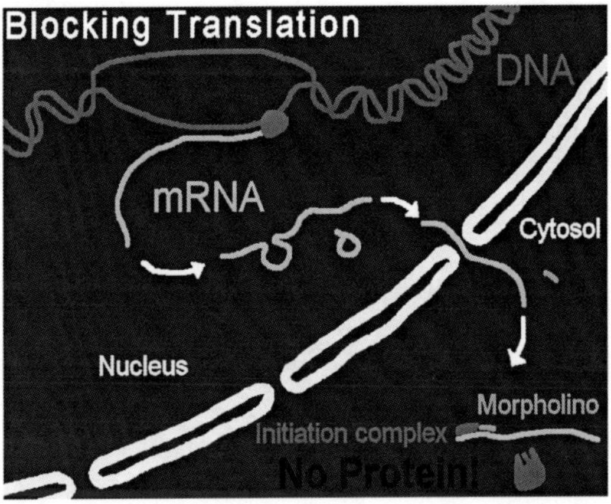

Figure: *Translation blocked by a Morpholino oligo*

Bound to the 5'-untranslated region of messenger RNA (mRNA), Morpholinos can interfere with progression of the ribosomal initiation complex from the 5' cap to the start codon. This prevents translation of the coding region of the targeted transcript (called "knocking down" gene expression). This is useful experimentally when an investigator wishes to know the function of a particular protein; Morpholinos provide a convenient means of knocking down expression of the protein and learning how that knockdown changes the cells or organism. Some Morpholinos knock down expression so effectively that, after degradation of preexisting proteins, the targeted proteins become undetectable by Western blot.

Modifying pre-mRNA Splicing

Morpholinos can interfere with pre-mRNA processing steps either by preventing splice-directing small nuclear ribonucleoproteins (snRNP) complexes from binding to their targets at the borders of introns on a strand of pre-mRNA, or by blocking the nucleophilic adenine base and preventing it from forming the splice lariat structure, or by interfering with the binding of splice regulatory proteins such as splice silencers and splice enhancers.

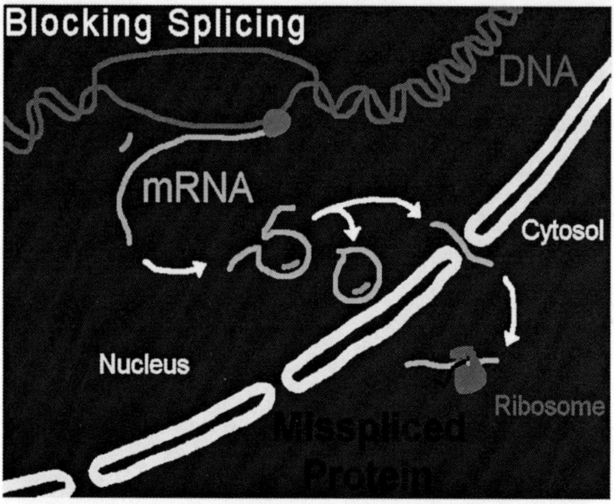

Figure: *Splicing blocked by a Morpholino oligo*

Preventing the binding of snRNP U1 (at the donor site) or U2/U5 (at the polypyrimidine moiety and acceptor site) can cause modified splicing, commonly excluding exons from the mature mRNA. Targeting some splice targets results in intron inclusions, while activation of cryptic splice sites can lead to partial inclusions or exclusions. Targets of U11/U12 snRNPs can also be blocked. Splice modification can be conveniently assayed by reverse-transcriptase polymerase chain reaction (RT-PCR) and is seen as a band shift after gel electrophoresis of RT-PCR products.

Blocking Other mRNA Sites

Morpholinos have been used to block miRNA activity and maturation. They can also block ribozyme activity. U2 and U12 snRNP functions have been inhibited by Morpholinos. Morpholinos targeted to "slippery" mRNA sequences within protein coding regions can induce translational frameshifts. Activities of Morpholinos against this variety of targets suggest that Morpholinos can be used as a general-purpose tool for blocking interactions of proteins or nucleic acids with mRNA.

Specificity, Stability and Non-antisense Effects

Morpholinos have become a standard knockdown tool in animal embryonic systems, which have a broader range of gene expression than adult cells and can be strongly affected by an off-target interaction. Following initial injections into frog or fish embryos at the single-cell or few-cell stages, Morpholino effects can be measured up to five days later, after most of the processes of organogenesis and differentiation are past, with observed phenotypes consistent with target-gene knockdown. Control oligos with irrelevant sequences usually produce no change in embryonic phenotype, evidence of the Morpholino oligo's sequence-specificity and lack of non-antisense effects. The dose required for a knockdown can be reduced by coinjection of several Morpholino oligos targeting the same mRNA, which is an effective strategy for reducing or eliminating dose-dependent off-target RNA interactions.

mRNA rescue experiments can often restore the wild-type phenotype to the embryos and provide evidence for the specificity of a Morpholino. In an mRNA rescue, a Morpholino is co-injected with an mRNA that codes for the same protein that the Morphlino is targeted to knock down. However, the rescue mRNA has a modified 5'-UTR (untranslated region) so that the rescue mRNA contains no target for the Morpholino but the rescue mRNA's coding region encodes the protein of interest. Translation of the rescue mRNA replaces production of the protein that was knocked down by the Morpholino. Since the rescue mRNA would not affect phenotypic changes due to modulation of off-target gene expression by the Morpholino, this return to wild-type phenotype is further evidence of Morpholino specificity.

Because of their completely unnatural backbones, Morpholinos are not recognized by cellular proteins. Nucleases do not degrade Morpholinos, nor are they degraded in serum or in cells. Morpholinos do not activate toll-like receptors and so they do not activate innate immune responses such as interferon induction or the NF-êB mediated inflammation response. Morpholinos are not known to modify methylation of DNA.

A cause for concern in the use of Morpholinos is the potential for "off-target" effects. Up to 18% of Morpholinos appear to have nontarget-related phenotypes including cell death in the central nervous system and somite tissues of zebrafish embryos. Most of these effects have been shown to be due to activation of p53-mediated apoptosis, and can be suppressed by co-injection of an anti-p53 Morpholino along with the experimental Morpholino; moreover, the p53-mediated apoptotic effect of a Morpholino knockdown has been phenocopied using another antisense structural type, showing the p53-mediated apoptosis to be a consequence of the loss of the targeted protein and not a consequence of the oligo type used to knock down the protein. It appears that these effects are sequence-specific; as in most cases, if a Morpholino is associated with non-target effects, the 4-base mismatch Morpholino will not trigger these effects.

The question of whether an observed morphant phenotype is due to the intended knockdown or an off-target interaction can often be addressed by running another experiment to confirm that the observed morphant phenotype results from the knockdown of the expected target. This can be done by recapitulating the morphant phenotype with a second, non-overlapping Morpholino targeting the same mRNA or by confirmation of the observed phenotypes by use of a mutant strain or dominant-negative methods. As mentioned above, rescue of observed phenotypes by coinjecting a rescue mRNA is, when feasible, a reliable test of specificity of a Morpholino.

Delivery

For a Morpholino to be effective, it must be delivered past the cell membrane into the cytosol of a cell. Once in the cytosol, Morpholinos freely diffuse between the cytosol and nucleus, as demonstrated by the nuclear splice-modifying activity of Morpholinos observed after microinjection into the cytosol of cells. Different methods are used for delivery into embryos, into cultured cells or into adult animals.

A microinjection apparatus is usually used for delivery into an embryo, with injections most commonly performed at the single-cell or few-cell stage; an alternative method for embryonic delivery is electroporation, which can deliver oligos into tissues of later embryonic stages. Common techniques for delivery into cultured cells include the Endo-Porter peptide (which causes the Morpholino to be released from endosomes), the Special Delivery system (using a Morpholino-DNA heteroduplex and an ethoxylated polyethylenimine delivery reagent), electroporation, or scrape loading.

Delivery into adult tissues is usually difficult, though there are a few systems allowing useful uptake of unmodified Morpholino oligos (including the inherently leaky muscle cells caused by Duchenne muscular dystrophy or the vascular endothelial cells stressed during balloon angioplasty).

Though they permeate through intercellular spaces in tissues effectively, unconjugated PMOs have limited distribution into the cytosol and nuclear spaces within healthy tissues following IV administration. Systemic delivery into many cells in adult organisms can be accomplished by using covalent conjugates of Morpholino oligos with cell-penetrating peptides, and, while toxicity has been associated with moderate doses of the peptide conjugates, they have been used in vivo for effective oligo delivery at doses below those causing observed toxicity. An octa-guanidinium dendrimer attached to the end of a Morpholino can deliver the modified oligo (called a Vivo-Morpholino) from the blood to the cytosol. Delivery-enabled Morpholinos, such as peptide conjugates and Vivo-Morpholinos, show promise as therapeutics for viral and genetic diseases.

Peptide Nucleic Acid

Peptide nucleic acid (PNA) is an artificially synthesized polymer similar to DNA or RNA invented by Peter E. Nielsen (Univ. Copenhagen), Michael

Egholm (Univ. Copenhagen), Rolf H. Berg (Risø National Lab), and Ole Buchardt (Univ. Copenhagen) in 1991. DNA and RNA have a deoxyribose and ribose sugar backbone, respectively, whereas PNA's backbone is composed of repeating N-(2-aminoethyl)-glycine units linked by peptide bonds. PNA is not known to occur naturally. The various purine and pyrimidine bases are linked to the backbone by methylene carbonyl bonds. PNAs are depicted like peptides, with the N-terminus at the first (left) position and the C-terminus at the right.

Since the backbone of PNA contains no charged phosphate groups, the binding between PNA/DNA strands is stronger than between DNA/DNA strands due to the lack of electrostatic repulsion. Unfortunately, this also causes it to be rather hydrophobic, which makes it difficult to deliver to body cells in solution without being flushed out of the body first. Early experiments with homopyrimidine strands (strands consisting of only one repeated pyrimidine base) have shown that the T_m ("melting" temperature) of a 6-base thymine PNA/adenine DNA double helix was 31 °C in comparison to an equivalent 6-base DNA/DNA duplex that denatures at a temperature less than 10 °C. Mixed base PNA molecules are true mimics of DNA molecules in terms of base-pair recognition. PNA/PNA binding is stronger than PNA/DNA binding.

Synthetic peptide nucleic acid oligomers have been used in recent years in molecular biology procedures, diagnostic assays and antisense therapies. Due to their higher binding strength it is not necessary to design long PNA oligomers for use in these roles, which usually require oligonucleotide probes of 20–25 bases. The main concern of the length of the PNA-oligomers is to guarantee the specificity. PNA oligomers also show greater specificity in binding to complementary DNAs, with a PNA/DNA base mismatch being more destabilizing than a similar mismatch in a DNA/DNA duplex. This binding strength and specificity also applies to PNA/RNA duplexes. PNAs are not easily recognized by either nucleases or proteases, making them resistant to enzyme degradation.

PNAs are also stable over a wide pH range. Though an unmodified PNA cannot readily cross cell membranes to enter the cytosol, covalently coupling a cell penetrating peptide to a PNA can improve cytosolic delivery.

It has been hypothesized that the earliest life on Earth may have used PNA as a genetic material due to its extreme robustness, simpler formation and possible spontaneous polymerization at 100°C (while water at standard pressure boils at this temperature, water at high pressure—as in deep ocean—boils at higher temperatures). If this is so, life evolved to a DNA/RNA-based system only at a later stage. Evidence for this PNA world hypothesis is however far from conclusive.

The name is somewhat of a misnomer as PNA is not an acid.

Applications

- Alteration of gene expression - both as inhibitor and promoter in different cases
- Antigene and antisense therapeutic agent
- Anticancer agent
- Antiviral, Antibacterial and Antiparasitic agent
- Molecular tools and probes of biosensor
- Detection of DNA sequences
- Locked Nucleic Acid

A locked nucleic acid (LNA), often referred to as inaccessible RNA, is a modified RNA nucleotide. The ribose moiety of an LNA nucleotide is modified with an extra bridge connecting the 2' oxygen and 4' carbon. The bridge "locks" the ribose in the 3'-endo (North) conformation, which is often found in the A-form duplexes. LNA nucleotides can be mixed with DNA or RNA residues in the oligonucleotide whenever desired. Such oligomers are synthesized chemically and are commercially available. The locked ribose conformation enhances base stacking and backbone pre-organization.

This significantly increases the hybridization properties (melting temperature) of oligonucleotides. LNA was independently synthesized by the group of Jesper Wengel in 1998, soon after the first synthesis by the group of Takeshi Imanishi in 1997. The exclusive rights to the LNA technology were secured in 1997 by Exiqon A/S, a Danish biotech company. LNA nucleotides are used to increase the sensitivity and specificity of expression in DNA microarrays, FISH probes, real-time PCR probes and other molecular biology techniques based on oligonucleotides. For the in situ detection of miRNA the use of LNA is currently (2005) the only efficient method.

A triplet of LNA nucleotides surrounding a single-base mismatch site maximizes LNA probe specificity unless the probe contains the guanine base of G-T mismatch. Using LNA based oligonucleotides therapeutically is an emerging field in biotechnology. The Danish pharmaceutical company Santaris Pharma a/s owns the sole rights to therapeutic uses of LNA technology, and is now developing a new, LNA based, hepatitis C drug called miravirsen, targeting miR-122, which is in Phase II clinical testing as of late 2010.

Benefits of the LNA Technology

Some of the benefits of using LNA include:

- Ideal for the detection of short RNA and DNA targets
- Increases the thermal stability of duplexes

- Capable of single nucleotide discrimination
- Resistant to exo- and endonucleases resulting in high stability in vivo and in vitro applications
- Increased target specificity
- Facilitates Tm normalization
- Strand invasion properties enables detection of "hard to access" samples
- Compatible with standard enzymatic processes

Applications of the LNA technology

Some proven applications of LNA include:

- Allele-specific PCR: allows for the design of shorter primers, without compromising binding specificity
- Microarray gene expression profiling: provides increased sensitivity and selectivity with smaller amounts of substrates
- Small RNA research
- SNP genotyping
- mRNA antisense oligonucleotides
- RNAi
- DNAzymes
- Fluorescence Polarization probes
- Molecular Beacons
- Gene repair/exon skipping
- Splice variant detection
- Comparative genome hybridization (GCH)

Other therapeutic and diagnostic applications of LNA technology are in development.

RNA Interference (which uses double-strand RNA)

RNA interference (RNAi) is a system within living cells that takes part in controlling which genes are active and how active they are. Two types of small RNA molecules – microRNA (miRNA) and small interfering RNA (siRNA) – are central to RNA interference. RNAs are the direct products of genes, and these small RNAs can bind to other specific RNAs (mRNA) and either increase or decrease their activity, for example by preventing a messenger RNA from producing a protein. RNA interference has an important role in defending cells against parasitic genes – viruses and transposons – but also in directing development as well as gene expression in general.

The RNAi pathway is found in many eukaryotes including animals and is initiated by the enzyme Dicer, which cleaves long double-stranded RNA (dsRNA) molecules into short fragments of ~20 nucleotides that are called siRNAs. Each siRNA is unwound into two single-stranded (ss) ssRNAs, namely the passenger strand and the guide strand. The passenger strand will be degraded, and the guide strand is incorporated into the RNA-induced silencing complex (RISC). The most well-studied outcome is post-transcriptional gene silencing, which occurs when the guide strand base pairs with a complementary sequence of a messenger RNA molecule and induces cleavage by Argonaute, the catalytic component of the RISC complex. This process is known to spread systemically throughout the organism despite initially limited molar concentrations of siRNA.

The selective and robust effect of RNAi on gene expression makes it a valuable research tool, both in cell culture and in living organisms because synthetic dsRNA introduced into cells can induce suppression of specific genes of interest. RNAi may also be used for large-scale screens that systematically shut down each gene in the cell, which can help identify the components necessary for a particular cellular process or an event such as cell division.

Exploitation of the pathway is also a promising tool in biotechnology and medicine. Historically, RNA interference was known by other names, including cosuppression, post transcriptional gene silencing, and quelling. Only after these apparently unrelated processes were fully understood did it become clear that they all described the RNAi phenomenon. In 2006, Andrew Fire and Craig C. Mello shared the Nobel Prize in Physiology or Medicine for their work on RNA interference in the nematode worm C. elegans, which they published in 1998.

Cellular Mechanism

The dicer protein from Giardia intestinalis, which catalyzes the cleavage of dsRNA to siRNAs. The RNase domains are colored green, the PAZ domain yellow, the platform domain red, and the connector helix blue. RNAi is an RNA-dependent gene silencing process that is controlled by the RNA-induced silencing complex (RISC) and is initiated by short double-stranded RNA molecules in a cell's cytoplasm, where they interact with the catalytic RISC component argonaute. When the dsRNA is exogenous (coming from infection by a virus with an RNA genome or laboratory manipulations), the RNA is imported directly into the cytoplasm and cleaved to short fragments by the enzyme. The initiating dsRNA can also be endogenous (originating in the cell), as in pre-microRNAs expressed from RNA-coding genes in the genome. The primary transcripts from such genes are first processed to form the characteristic stem-loop structure of pre-miRNA in the nucleus, then exported to the cytoplasm to be cleaved by Dicer. Thus, the two dsRNA pathways, exogenous and endogenous, converge at the RISC complex.

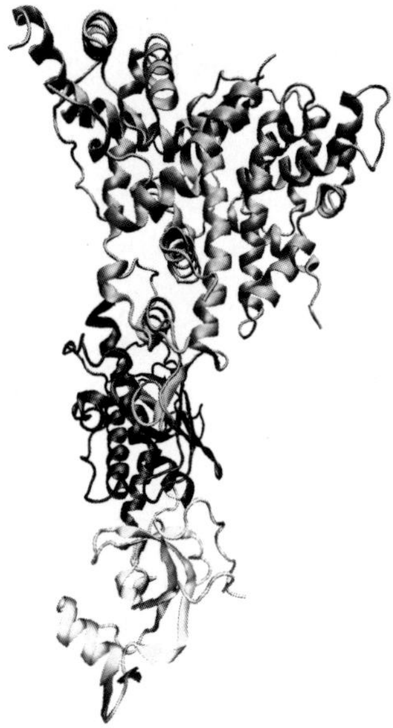

dsRNA Cleavage

Endogenous dsRNA initiates RNAi by activating the ribonuclease protein Dicer, which binds and cleaves double-stranded RNAs (dsRNAs) to produce double-stranded fragments of 20–25 base pairs with a 2-nucleotide overhang at the 3' end. Bioinformatics studies on the genomes of multiple organisms suggest this length maximizes target-gene specificity and minimizes non-specific effects. These short double-stranded fragments are called small interfering RNAs (siRNAs). These siRNAs are then separated into single strands and integrated into an active RISC complex. After integration into the RISC, siRNAs base-pair to their target mRNA and induce cleavage of the mRNA, thereby preventing it from being used as a translation template.

Exogenous dsRNA is detected and bound by an effector protein, known as RDE-4 in C. elegans and R2D2 in Drosophila, that stimulates dicer activity. This protein only binds long dsRNAs, but the mechanism producing this length specificity is unknown. This RNA-binding protein then facilitates the transfer of cleaved siRNAs to the RISC complex.

In C. elegans, this initiation response is amplified through the synthesis of a population of 'secondary' siRNAs during which the dicer-produced initiating or 'primary' siRNAs are used as templates. These 'secondary' siRNAs are structurally distinct from dicer-produced siRNAs and appear to be produced by an RNA-dependent RNA polymerase (RdRP).

MicroRNA

MicroRNAs (miRNAs) are genomically encoded non-coding RNAs that help regulate gene expression, particularly during development. The phenomenon of RNA interference, broadly defined, includes the endogenously induced gene silencing effects of miRNAs as well as silencing triggered by foreign dsRNA. Mature miRNAs are structurally similar to siRNAs produced from exogenous dsRNA, but before reaching maturity, miRNAs must first undergo extensive post-transcriptional modification.

Figure: *The stem-loop secondary structure of a pre-microRNA from Brassica oleracea.*

An miRNA is expressed from a much longer RNA-coding gene as a primary transcript known as a pri-miRNA which is processed, in the cell nucleus, to a 70-nucleotide stem-loop structure called a pre-miRNA by the microprocessor complex. This complex consists of an RNase III enzyme called Drosha and a dsRNA-binding protein Pasha. The dsRNA portion of this pre-miRNA is bound and cleaved by Dicer to produce the mature miRNA molecule that can be integrated into the RISC complex; thus, miRNA and siRNA share the same cellular machinery downstream of their initial processing.

The siRNAs derived from long dsRNA precursors differ from miRNAs in that miRNAs, especially those in animals, typically have incomplete base pairing to a target and inhibit the translation of many different mRNAs with similar sequences. In contrast, siRNAs typically base-pair perfectly and induce mRNA cleavage only in a single, specific target. In Drosophila and C. elegans, miRNA and siRNA are processed by distinct argonaute proteins and dicer enzymes.

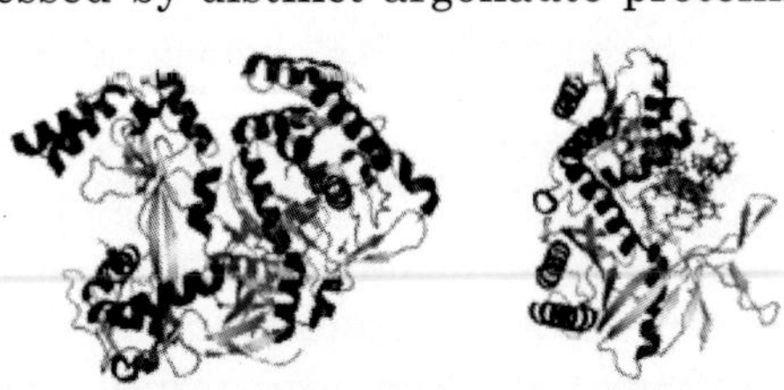

Figure: *Left: A full-length argonaute protein from the archaea species Pyrococcus furiosus. Right: The PIWI domain of an argonaute protein in complex with double-stranded RNA.*

RISC Activation and Catalysis

The active components of an RNA-induced silencing complex (RISC) are endonucleases called argonaute proteins, which cleave the target mRNA strand complementary to their bound siRNA. As the fragments produced by dicer are double-stranded, they could each in theory produce a functional siRNA. However, only one of the two strands, which is known as the guide strand, binds the argonaute protein and directs gene silencing.

The other anti-guide strand or passenger strand is degraded during RISC activation. Although it was first believed that an ATP-dependent helicase separated these two strands, the process is actually ATP-independent and performed directly by the protein components of RISC. The strand selected as the guide tends to be the one whose 5' end is less stably paired to its complement, but strand selection is unaffected by the direction in which dicer cleaves the dsRNA before RISC incorporation. Instead, the R2D2 protein may serve as the differentiating factor by binding the more-stable 5' end of the passenger strand.

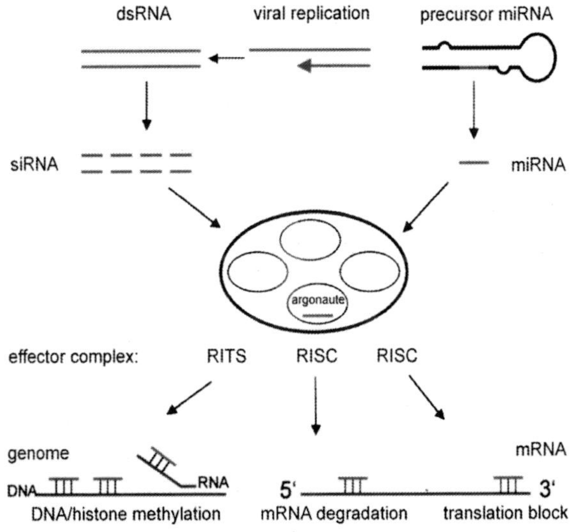

The structural basis for binding of RNA to the argonaute protein was examined by X-ray crystallography of the binding domain of an RNA-bound argonaute protein. Here, the phosphorylated 5' end of the RNA strand enters a conserved basic surface pocket and makes contacts through a divalent cation (an atom with two positive charges) such as magnesium and by aromatic stacking (a process that allows more than one atom to share an electron by passing it back and forth) between the 5' nucleotide in the siRNA and a conserved tyrosine residue. This site is thought to form a nucleation site for the binding of the siRNA to its mRNA target.

It is not understood how the activated RISC complex locates complementary mRNAs within the cell. Although the cleavage process has been proposed to be linked to translation, translation of the mRNA target is not essential for RNAi-mediated degradation. Indeed, RNAi may be more effective against mRNA targets that are not translated. Argonaute proteins, the catalytic components of RISC, are localized to specific regions in the cytoplasm called P-bodies (also cytoplasmic bodies or GW bodies), which are regions with high rates of mRNA decay; miRNA activity is also clustered in P-bodies. Disruption of P-bodies decreases the efficiency of RNA interference, suggesting that they are the site of a critical step in the RNAi process.

The enzyme dicer trims double stranded RNA, to form small interfering RNA or microRNA. These processed RNAs are incorporated into the RNA-induced silencing complex (RISC), which targets messenger RNA to prevent translation.

Transcriptional Silencing

Components of the RNA interference pathway are also used in many eukaryotes in the maintenance of the organization and structure of their genomes. Modification of histones and associated induction of heterochromatin formation serves to downregulate genes pre-transcriptionally; this process is referred to as RNA-induced transcriptional silencing (RITS), and is carried out by a complex of proteins called the RITS complex. In fission yeast this complex contains argonaute, a chromodomain protein Chp1, and a protein called Tas3 of unknown function. As a consequence, the induction and spread of heterochromatic regions requires the argonaute and RdRP proteins. Indeed, deletion of these genes in the fission yeast S. pombe disrupts histone methylation and centromere formation, causing slow or stalled anaphase during cell division. In some cases, similar processes associated with histone modification have been observed to transcriptionally upregulate genes.

The mechanism by which the RITS complex induces heterochromatin formation and organization is not well understood, and most studies have focused on the mating-type region in fission yeast, which may not be representative of activities in other genomic regions or organisms. In maintenance of existing heterochromatin regions, RITS forms a complex with siRNAs complementary to the local genes and stably binds local methylated histones, acting co-transcriptionally to degrade any nascent pre-mRNA transcripts that are initiated by RNA polymerase. The formation of such a heterochromatin region, though not its maintenance, is dicer-dependent, presumably because dicer is required to generate the initial complement of siRNAs that target subsequent transcripts. Heterochromatin maintenance has been suggested to function as a self-reinforcing feedback loop, as new siRNAs are formed from the occasional nascent transcripts by RdRP for incorporation into local RITS complexes. The relevance of observations from fission yeast mating-type regions and centromeres to mammals is not clear, as heterochromatin maintenance in mammalian cells may be independent of the components of the RNAi pathway.

Crosstalk with RNA Editing

The type of RNA editing that is most prevalent in higher eukaryotes converts adenosine nucleotides into inosine in dsRNAs via the enzyme adenosine deaminase (ADAR). It was originally proposed in 2000 that the RNAi and A'!I RNA editing pathways might compete for a common dsRNA substrate. Indeed, some pre-miRNAs do undergo A'!I RNA editing, and this mechanism may regulate the processing and expression of mature miRNAs. Furthermore, at least one mammalian ADAR can sequester siRNAs from RNAi pathway components. Further support for this model comes from studies on ADAR-null C. elegans strains indicating that A'!I RNA editing may counteract RNAi silencing of endogenous genes and transgenes.

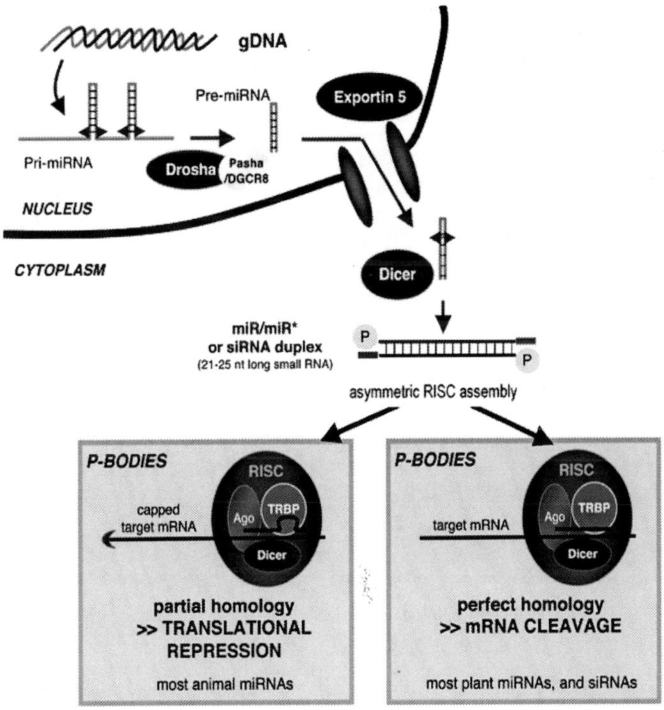

Figure: *Illustration of the major differences between plant and animal gene silencing. Natively expressed microRNA or exogenous small interfering RNA is processed by dicer and integrated into the RISC complex, which mediates gene silencing.*

Variation Among Organisms

Organisms vary in their ability to take up foreign dsRNA and use it in the RNAi pathway. The effects of RNA interference can be both systemic and heritable in plants and C. elegans, although not in Drosophila or mammals.

In plants, RNAi is thought to propagate by the transfer of siRNAs between cells through plasmodesmata (channels in the cell walls that enable communication and transport). The heritability comes from methylation of promoters targeted by RNAi; the new methylation pattern is copied in each new generation of the cell. A broad general distinction between plants and animals lies in the targeting of endogenously produced miRNAs; in plants, miRNAs are usually perfectly or nearly perfectly complementary to their target genes and induce direct mRNA cleavage by RISC, while animals' miRNAs tend to be more divergent in sequence and induce translational repression. This translational effect may be produced by inhibiting the interactions of translation initiation factors with the messenger RNA's polyadenine tail.

Some eukaryotic protozoa such as Leishmania major and Trypanosoma cruzi lack the RNAi pathway entirely. Most or all of the components are also missing in some fungi, most notably the model organism Saccharomyces cerevisiae. A recent study however reveals the presence of RNAi in other budding yeast species such as Saccharomyces castellii and Candida albicans, further demonstrating that inducing two RNAi-related proteins from S. castellii facilitates RNAi in S. cerevisiae. That certain ascomycetes and basidiomycetes are missing RNA interference pathways indicates that proteins required for RNA silencing have been lost independently from many fungal lineages, possibly due to the evolution of a novel pathway with similar function, or to the lack of selective advantage in certain niches.

Related Prokaryotic Systems

Gene expression in prokaryotes is influenced by an RNA-based system similar in some respects to RNAi. Here, RNA-encoding genes control mRNA abundance or translation by producing a complementary RNA that binds to an mRNA by base pairing. However these regulatory RNAs are not generally considered to be analogous to miRNAs because the dicer enzyme is not involved. It has been suggested that CRISPR interference systems in prokaryotes are analogous to eukaryotic RNA interference systems, although none of the protein components are orthologous.

Biological Functions

Immunity

RNA interference is a vital part of the immune response to viruses and other foreign genetic material, especially in plants where it may also prevent the self-propagation of transposons. Plants such as Arabidopsis thaliana express multiple dicer homologs that are specialized to react differently when the plant is exposed to different types of viruses. Even before the RNAi pathway was fully understood, it was known that induced gene silencing in plants could spread throughout the plant in a systemic effect, and could be transferred from stock to scion plants via grafting.

This phenomenon has since been recognized as a feature of the plant adaptive immune system, and allows the entire plant to respond to a virus after an initial localized encounter. In response, many plant viruses have evolved elaborate mechanisms that suppress the RNAi response in plant cells. These include viral proteins that bind short double-stranded RNA fragments with single-stranded overhang ends, such as those produced by the action of dicer. Some plant genomes also express endogenous siRNAs in response to infection by specific types of bacteria. These effects may be part of a generalized response to pathogens that downregulates any metabolic processes in the host that aid the infection process.

Although animals generally express fewer variants of the dicer enzyme than plants, RNAi in some animals has also been shown to produce an antiviral response.

In both juvenile and adult Drosophila, RNA interference is important in antiviral innate immunity and is active against pathogens such as Drosophila X virus. A similar role in immunity may operate in C. elegans, as argonaute proteins are upregulated in response to viruses and worms that overexpress components of the RNAi pathway are resistant to viral infection.

Downregulation of Genes

Endogenously expressed miRNAs, including both intronic and intergenic miRNAs, are most important in translational repression and in the regulation of development, especially on the timing of morphogenesis and the maintenance of undifferentiated or incompletely differentiated cell types such as stem cells. The role of endogenously expressed miRNA in downregulating gene expression was first described in C. elegans in 1993. In plants this function was discovered when the "JAW microRNA" of Arabidopsis was shown to be involved in the regulation of several genes that control plant shape. In plants, the majority of genes regulated by miRNAs are transcription factors; thus miRNA activity is particularly wide-ranging and regulates entire gene networks during development by modulating the expression of key regulatory genes, including transcription factors as well as F-box proteins. In many organisms, including humans, miRNAs have also been linked to the formation of tumors and dysregulation of the cell cycle. Here, miRNAs can function as both oncogenes and tumor suppressors.

Upregulation of Genes

RNA sequences (siRNA and miRNA) that are complementary to parts of a promoter can increase gene transcription, a phenomenon dubbed RNA activation. Part of the mechanism for how these RNA upregulate genes is known: dicer and argonaute are involved, possibly via histone demethylation.. miRNAs have also been proposed to upregulate their target genes upon cell cycle arrest, although the underlying mechanisms have not been elucidated.

Evolution

Based on parsimony-based phylogenetic analysis, the most recent common ancestor of all eukaryotes most likely already possessed an early RNA interference pathway; the absence of the pathway in certain eukaryotes is thought to be a derived characteristic. This ancestral RNAi system probably contained at least one dicer-like protein, one argonaute, one PIWI protein, and an RNA-dependent RNA polymerase that may have also played other cellular roles. A large-scale comparative genomics study likewise indicates that the eukaryotic crown group already possessed these components, which may then have had closer functional associations with generalized RNA degradation systems such as the exosome.

This study also suggests that the RNA-binding argonaute protein family, which is shared among eukaryotes, most archaea, and at least some bacteria (such as Aquifex aeolicus), is homologous to and originally evolved from components of the translation initiation system.

The ancestral function of the RNAi system is generally agreed to have been immune defense against exogenous genetic elements such as transposons and viral genomes. Related functions such as histone modification may have already been present in the ancestor of modern eukaryotes, although other functions such as regulation of development by miRNA are thought to have evolved later.

RNA interference genes, as components of the antiviral innate immune system in many eukaryotes, are involved in an evolutionary arms race with viral genes. Some viruses have evolved mechanisms for suppressing the RNAi response in their host cells, an effect that has been noted particularly for plant viruses. Studies of evolutionary rates in Drosophila have shown that genes in the RNAi pathway are subject to strong directional selection and are among the fastest-evolving genes in the Drosophila genome.

3

The Basic Microarray

Since an array can contain tens of thousands of probes, a microarray experiment can accomplish many genetic tests in parallel. Therefore arrays have dramatically accelerated many types of investigation. In standard microarrays, the probes are synthesized and then attached via surface engineering to a solid surface by a covalent bond to a chemical matrix (via epoxy-silane, amino-silane, lysine, polyacrylamide or others).

The solid surface can be glass or a silicon chip, in which case they are colloquially known as an Affy chip when an Affymetrix chip is used. Other microarray platforms, such as Illumina, use microscopic beads, instead of the large solid support. Alternatively, microarrays can be constructed by the direct synthesis of oligonucleotide probes on solid surfaces. DNA arrays are different from other types of microarray only in that they either measure DNA or use DNA as part of its detection system.

DNA microarrays can be used to measure changes in expression levels, to detect single nucleotide polymorphisms (SNPs), or to genotype or resequence mutant genomes. Microarrays also differ in fabrication, workings, accuracy, efficiency, and cost. Additional factors for microarray experiments are the experimental design and the methods of analyzing the data.

History

Microarray technology evolved from Southern blotting, where fragmented DNA is attached to a substrate and then probed with a known gene or fragment. The first reported use of this approach was the analysis of 378 arrayed lysed bacterial colonies each harboring a different sequence which were assayed in multiple replicas for expression of the genes in multiple normal and tumor tissue. This was expanded to analysis of more than 4000 human sequences with computer driven scanning and image processing for quantitative analysis of the sequences in human colonic tumors and normal tissue and then to comparison of colonic tissues at different genetic risk. The use of a collection of distinct DNAs in arrays for expression profiling was also described in 1987, and the arrayed DNAs were used to identify genes whose expression is modulated by interferon. These early gene arrays were made by spotting cDNAs onto filter paper with a pin-spotting device. The use of miniaturized microarrays for gene expression profiling was first reported in 1995, and a complete eukaryotic genome (Saccharomyces cerevisiae) on a microarray was published in 1997.

Principle

The core principle behind microarrays is hybridization between two DNA strands, the property of complementary nucleic acid sequences to specifically pair with each other by forming hydrogen bonds between complementary nucleotide base pairs.

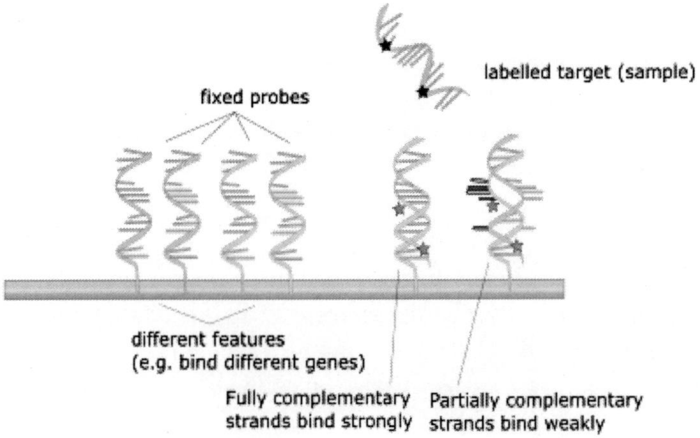

Figure: *Hybridization of the target to the probe.*

A high number of complementary base pairs in a nucleotide sequence means tighter non-covalent bonding between the two strands. After washing off of non-specific bonding sequences, only strongly paired strands will remain hybridized. So fluorescently labeled target sequences that bind to a probe sequence generate a signal that depends on the strength of the hybridization determined by the number of paired bases, the hybridization conditions (such as temperature), and washing after hybridization. Total strength of the signal, from a spot (feature), depends upon the amount of target sample binding to the probes present on that spot. Microarrays use relative quantization in which the intensity of a feature is compared to the intensity of the same feature under a different condition, and the identity of the feature is known by its position.

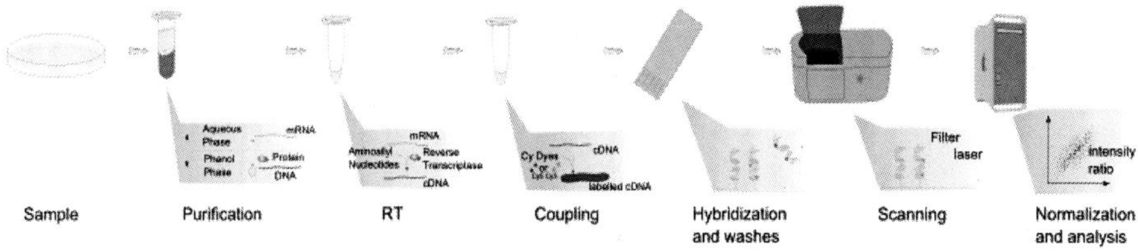

Figure: *The step required in a microarray experiment.*

Uses and Types

Figure: *Two Affymetrix chips. A match is shown at bottom left for size comparison.*

Many types of array exist and the broadest distinction is whether they are spatially arranged on a surface or on coded beads:

- The traditional solid-phase array is a collection of orderly microscopic "spots", called features, each with a specific probe attached to a solid surface, such as glass, plastic or silicon biochip (commonly known as a genome chip, DNA chip or gene array). Thousands of them can be placed in known locations on a single DNA microarray.
- The alternative bead array is a collection of microscopic polystyrene beads, each with a specific probe and a ratio of two or more dyes, which do not interfere with the fluorescent dyes used on the target sequence.

DNA microarrays can be used to detect DNA (as in comparative genomic hybridization), or detect RNA (most commonly as cDNA after reverse transcription) that may or may not be translated into proteins. The process of measuring gene expression via cDNA is called expression analysis or expression profiling.

***Table:** Applications include*

Application or technology	***Synopsis***
Gene expression profiling	In an mRNA or gene expression profiling experiment the expression levels of thousands of genes are simultaneously monitored to study the effects of certain treatments, diseases, and developmental stages on gene expression. For example, microarray-based gene expression profiling can be used to identify genes whose expression is changed in response to pathogens or other organisms by comparing gene expression in infected to that in uninfected cells or tissues.
Comparative geno -mic hybridization	Assessing genome content in different cells or closely related organisms.
GeneID	Small microarrays to check IDs of organisms in food and feed (like GMO), mycoplasms in cell culture, or pathogens for disease detection, mostly combining PCR and microarray technology.
Chromatin immunop-recipitation on Chipby	DNA sequences bound to a particular protein can be isolated immunoprecipitating that protein (ChIP), these fragments can be then hybridized to a microarray (such as a tiling array) allowing the determination of protein binding site occupancy throughout the genome. Example protein to immunoprecipitate are histone modifications (H3K27me3, H3K4me2, H3K9me3, etc.), Polycomb-group protein (PRC2:Suz12, PRC1:YY1) and trithorax-group protein (Ash1) to study the epigenetic landscape or RNA Polymerase II to study the transcription landscape.
DamID	Analogously to ChIP, genomic regions bound by a protein of interest can be isolated and used to probe a microarray to determine binding site occupancy. Unlike ChIP, DamID does not require antibodies but makes use of adenine methylation near the protein's binding sites to selectively amplify those regions, introduced by expressing minute amounts of protein of interest fused to bacterial DNA adenine methyltransferase.
SNP detection	Identifying single nucleotide polymorphism among alleles within or between populations. Several applications of microarrays make use of SNP detection, including Genotyping, forensic analysis, measuring predisposition to disease, identifying drug-candidates, evaluating germline mutations in individuals or somatic mutations in cancers, assessing loss of heterozygosity, or genetic linkage analysis.
Alternative splicing	detection An 'exon junction array design uses probes specific to the expected or potential splice sites of predicted exons for a gene. It is of intermediate density, or coverage, to a typical gene expression array (with 1-3 probes per gene) and a genomic tiling array (with hundreds or thousands of probes per gene). It is used to assay the expression of alternative splice forms of a gene. Exon arrays have a different design, employing probes designed to detect each individual exon for known or predicted genes, and can be used for detecting different splicing isoforms.
Fusion genes microarray	A Fusion gene microarray can detect fusion transcripts, e.g. from cancer specimens. The principle behind this is building on the alternative splicing microarrays. The oligo design strategy enables combined measurements of chimeric transcript junctions with exon-wise measurements of individual fusion partners.

Tiling array	Genome tiling arrays consist of overlapping probes designed to densely represent a genomic region of interest, sometimes as large as an entire human chromosome. The purpose is to empirically detect expression of transcripts or alternatively splice forms which may not have been previously known or predicted.

Fabrication

Microarrays can be manufactured in different ways, depending on the number of probes under examination, costs, customization requirements, and the type of scientific question being asked. Arrays may have as few as 10 probes or up to 2.1 million micrometre-scale probes from commercial vendors.

Microarrays can be fabricated using a variety of technologies, including printing with fine-pointed pins onto glass slides, photolithography using pre-made masks, photolithography using dynamic micromirror devices, ink-jet printing, or electrochemistry on microelectrode arrays.

In spotted microarrays, the probes are oligonucleotides, cDNA or small fragments of PCR products that correspond to mRNAs. The probes are synthesized prior to deposition on the array surface and are then "spotted" onto glass. A common approach utilizes an array of fine pins or needles controlled by a robotic arm that is dipped into wells containing DNA probes and then depositing each probe at designated locations on the array surface. The resulting "grid" of probes represents the nucleic acid profiles of the prepared probes and is ready to receive complementary cDNA or cRNA "targets" derived from experimental or clinical samples.

This technique is used by research scientists around the world to produce "in-house" printed microarrays from their own labs. These arrays may be easily customized for each experiment, because researchers can choose the probes and printing locations on the arrays, synthesize the probes in their own lab (or collaborating facility), and spot the arrays. They can then generate their own labeled samples for hybridization, hybridize the samples to the array, and finally scan the arrays with their own equipment. This provides a relatively low-cost microarray that may be customized for each study, and avoids the costs of purchasing often more expensive commercial arrays that may represent vast numbers of genes that are not of interest to the investigator. Publications exist which indicate in-house spotted microarrays may not provide the same level of sensitivity compared to commercial oligonucleotide arrays, possibly owing to the small batch sizes and reduced printing efficiencies when compared to industrial manufactures of oligo arrays.

In oligonucleotide microarrays, the probes are short sequences designed to match parts of the sequence of known or predicted open reading frames. Although oligonucleotide probes are often used in "spotted" microarrays, the term "oligonucleotide array" most often refers to a specific technique of manufacturing. Oligonucleotide arrays are produced by printing short oligonucleotide sequences designed to represent a single gene or family of gene splice-variants by synthesizing this sequence directly onto the array surface instead of depositing intact sequences.

Sequences may be longer (60-mer probes such as the Agilent design) or shorter (25-mer probes produced by Affymetrix) depending on the desired purpose; longer probes are more specific to individual target genes, shorter probes may be spotted in higher density across the array and are cheaper to manufacture. One technique used to produce oligonucleotide arrays include photolithographic synthesis (Affymetrix) on a silica substrate where light and light-sensitive masking agents are used to "build" a sequence one nucleotide at a time across the entire array. Each applicable probe is selectively "unmasked" prior to bathing the array in a solution of a single nucleotide, then a masking reaction takes place and the next set of probes are unmasked in preparation for a different nucleotide exposure. After many repetitions, the sequences of every probe become fully constructed. More recently, Maskless Array Synthesis from NimbleGen Systems has combined flexibility with large numbers of probes.

Two-channel vs. One-channel detection

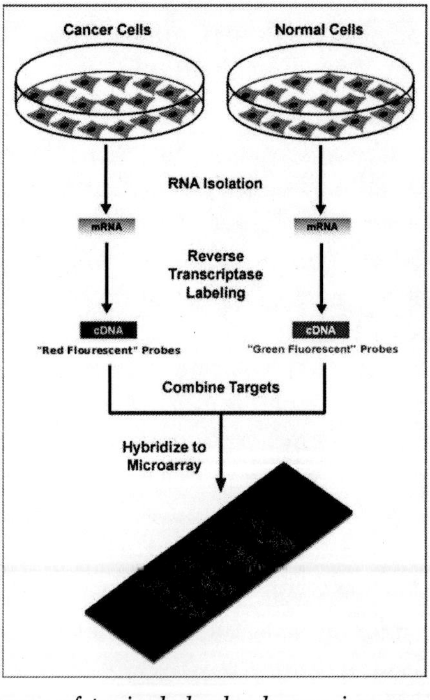

Figure: *Diagram of typical dual-colour microarray experiment.*

Two-color microarrays or two-channel microarrays are typically hybridized with cDNA prepared from two samples to be compared (e.g. diseased tissue versus healthy tissue) and that are labeled with two different fluorophores. Fluorescent dyes commonly used for cDNA labeling include Cy3, which has a fluorescence emission wavelength of 570 nm (corresponding to the green part of the light spectrum), and Cy5 with a fluorescence emission wavelength of 670 nm (corresponding to the red part ofm the light spectrumm).

The two Cy-labeled cDNA samples are mixed and hybridized to a single microarray that is then scanned in a microarray scanner to visualize fluorescence of the two fluorophores after excitation with a laser beam of a defined wavelength. Relative intensities of each fluorophore may then be used in ratio-based analysis to identify up-regulated and down-regulated genes.

Oligonucleotide microarrays often carry control probes designed to hybridize with RNA spike-ins. The degree of hybridization between the spike-ins and the control probes is used to normalize the hybridization measurements for the target probes. Although absolute levels of gene expression may be determined in the two-color array in rare instances, the relative differences in expression among different spots within a sample and between samples is the preferred method of data analysis for the two-color system. Examples of providers for such microarrays includes Agilent with their Dual-Mode platform, Eppendorf with their DualChip platform for colorimetric Silverquant labeling, and TeleChem International with Arrayit.

In single-channel microarrays or one-color microarrays, the arrays provide intensity data for each probe or probe set indicating a relative level of hybridization with the labeled target. However, they do not truly indicate abundance levels of a gene but rather relative abundance when compared to other samples or conditions when processed in the same experiment. Each RNA molecule encounters protocol and batch-specific bias during amplification, labeling, and hybridization phases of the experiment making comparisons between genes for the same microarray uninformative.

The comparison of two conditions for the same gene requires two separate single-dye hybridizations. Several popular single-channel systems are the Affymetrix "Gene Chip", Illumina "Bead Chip", Agilent single-channel arrays, the Applied Microarrays "CodeLink" arrays, amnd the Eppendorf "DualChip & Silverquant". One strength of the single-dye system lies in the fact that an aberrant sample cannot affect the raw data derived from other samples, because each array chip is exposed to only one sample (as opposed to a two-color system in which a single low-quality sample may drastically impinge on overall data precision even if the other sample was of high quality). Another benefit is that data are more easily compared to arrays from different experiments so long as batch effects have been accounted for. A drawback to the one-color system is that, when compared to the two-color system, twice as many microarrays are needed to compare samples within an experiment.

Microarrays and Bioinformatics

The advent of inexpensive microarray experiments created several specific bioinformatics challenges:

- the multiple levels of replication in experimental design (Experimental design)
- the number of platforms and independent groups and data format (Standardization)
- the treatment of the data (Statistical analysis)
- accuracy and precision (Relation between probe and gene)
- the sheer volume of data and the ability to share it (Data warehousing)

Experimental Design

Due to the biological complexity of gene expression, the considerations of experimental design that are discussed in the expression profiling article are of critical importance if statistically and biologically valid conclusions are to be drawn from the data.

There are three main elements to consider when designing a microarray experiment. First, replication of the biological samples is essential for drawing conclusions from the experiment. Second, technical replicates (two RNA samples obtained from each experimental unit) help to ensure precision and allow for testing differences within treatment groups.

The biological replicates include independent RNA extractions and technical replicates may be two aliquots of the same extraction. Third, spots of each cDNA clone or oligonucleotide are present as replicates (at least duplicates) on the microarray slide, to provide a measure of technical precision in each hybridization.

It is critical that information about the sample preparation and handling is discussed, in order to help identify the independent units in the experiment and to avoid inflated estimates of statistical significance.

Standardization

Microarray data is difficult to exchange due to the lack of standardization in platform fabrication, assay protocols, and analysis methods. This presents an interoperability problem in bioinformatics. Various grass-roots open-source projects are trying to ease the exchange and analysis of data produced with non-proprietary chips:

- For example, the "Minimum Information About a Microarray Experiment" (MIAME) checklist helps define the level of detail that should exist and is being adopted by many journals as a requirement for the submission of papers incorporating microarray results. But MIAME does not describe the format for the information, so while many formats can support the MIAME requirements, as of 2007 no format permits verification of complete semantic compliance.
- The "MicroArray Quality Control (MAQC) Project" is being conducted by the US Food and Drug Administration (FDA) to develop standards and quality control metrics which will eventually allow the use of MicroArray data in drug discovery, clinical practice and regulatory decision-making.
- The MGED Society has developed standards for the representation of gene expression experiment results and relevant annotations.

Statistical Analysis

Microarray data sets are commonly very large, and analytical precision is influenced by a number of variables. Statistical challenges include taking into account effects of background noise and appropriate normalization of the data.

Normalization methods may be suited to specific platforms and, in the case of commercial platforms, the analysis may be proprietary. Algorithms that affect statistical analysis include:

- Image analysis: gridding, spot recognition of the scanned image (segmentation algorithm), removal or marking of poor-quality and low-intensity features (called flagging).
- Data processing: background subtraction (based on global or local background), determination of spot intensities and intensity ratios, visualisation of data (e.g. see MA plot), and log-transformation of ratios, global or local normalization of intensity ratios, and segmentation into different copy number regions using step detection algorithms.
- Identification of statistically significant changes: t-test, ANOVA, Bayesian method Mann–Whitney test methods tailored to microarray data sets, which take into account multiple comparisons or cluster analysis. These methods assess statistical power based on the variation present in the data and the number of experimental replicates, and can help minimize Type I and type II errors in the analyses.
- Network-based methods: Statistical methods that take the underlying structure of gene networks into account, representing either associative or causative interactions or dependencies among gene products.

Microarray data may require further processing aimed at reducing the dimensionality of the data to aid comprehension and more focused analysis. Other methods permit analysis of data consisting of a low number of biological or technical replicates; for example, the Local Pooled Error (LPE) test pools standard deviations of genes with similar expression levels in an effort to compensate for insufficient replication.

Relation Between Probe and Gene

The relation between a probe and the mRNA that it is expected to detect is not trivial. Some mRNAs may cross-hybridize probes in the array that are supposed to detect another mRNA. In addition, mRNAs may experience amplification bias that is sequence or molecule-specific.

Thirdly, probes that are designed to detect the mRNA of a particular gene may be relying on genomic EST information that is incorrectly associated with that gene.

Data Warehousing

Microarray data was found to be more useful when compared to other similar datasets. The sheer volume of data, specialized formats (such as MIAME), and curation efforts associated with the datasets require specialized databases to store the data.

- Cyanine dyes, such as Cy3 and Cy5, are commonly used fluorophores with microarrays

Cyanine is a non-systematic name of a synthetic dye family belonging to polymethine group. Cyanines have many uses as fluorescent dyes, particularly in biomedical imaging. Depending on the structure, they cover the spectrum from IR to UV.

Cyanines were originally used, and still are, to increase the sensitivity range of photographic emulsions, i.e., to increase the range of wavelengths which will form an image on the film. Cyanines are also used in CD-R and DVD-R media.

The ones used are mostly green or light blue in color, and are chemically unstable. This makes unstabilized cyanine discs unsuitable for archival CD and DVD use, as they can fade and become unreadable in a few years, however, recent cyanine discs contain stabilizers that slow down the deterioration significantly.

These discs are often rated with an archival life of 75 years or more. The other dyes used in CD-Rs are phthalocyanine and azo.

Cyanines were first synthesized over a century ago, and there are a large number reported in the literature.

Structure

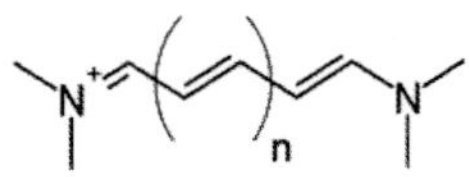

I

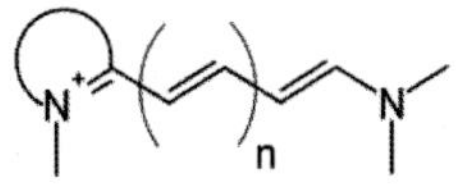

II

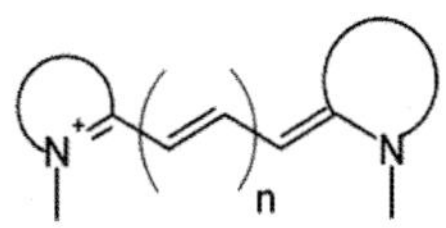

III

The Cyanines:

I = Streptocyanines,

II = Hemicyanines,

III = Closed cyanine

There are three types of cyanines:

- Streptocyanines or open chain cyanines:

R_2N =CH[CH=CH]$_n$-NR_2 (I)

- Hemicyanines:

Aryl=N =CH[CH=CH]$_n$-NR_2 (II)

- Closed chain cyanines:

Aryl=N =CH[CH=CH]$_n$-N=Aryl (III)

where two nitrogens are joined by a polymethine chain. Both nitrogens are each independently part of a heteroaromatic moiety, such as pyrrole, imidazole, thiazole, pyridine, quinoline, indole, benzothiazole, etc.

Cy3 and Cy5

Cy3 and Cy5 are reactive water-soluble fluorescent dyes of the cyanine dye family. Cy3 dyes are red (~550 nm excitation, ~570 nm emission and therefore appear red), while Cy5 is fluorescent in the far red region (~650/670 nm) but absorbs in the orange region (~649 nm). They are usually synthesized with reactive groups on either one or both of the nitrogen side chains so that they can be chemically linked to either nucleic acids or protein molecules. Labeling is done for visualization and quantification purposes. They are used in a wide variety of biological applications including comparative genomic hybridization and in gene chips, which are used in transcriptomics. They are also used to label proteins and nucleic acid for various studies including proteomics and RNA localization.

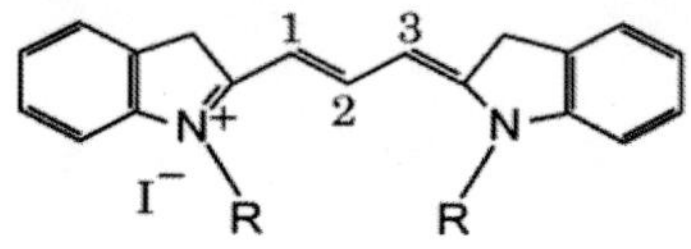

Cy3

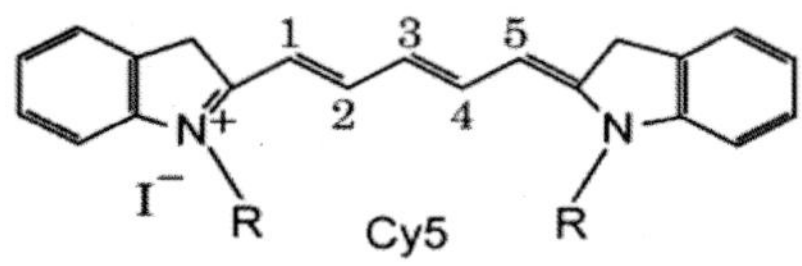

Nomenclature and Structure

Standard chemical names specify exactly the chemical structure of the molecule. The Cy3 and Cy5 nomenclature was first proposed by Ernst, et al. in 1989, and is non-standard, since it gives no hint of their chemical structures. In the original paper the number designated the count of the methines (as shown), and the side chains were unspecified. Thus various structures are designated Cy3 and Cy5 in the literature. The R groups do not have to be identical. In the dyes as used they are short aliphatic chains one or both of which ends in a highly reactive moieties such as N-hydroxysuccinimide or maleimide.

Table: *Spectral characteristics*

Dye	*Absorbance Max*	*Emission Max in PBS buffer*	*Quantum yield*	*Molecular weight (Da)*
Cy3	550 nm	570 nm	0.04	766
Cy5	649 nm	670 nm	0.28	792

The scanners actually use different laser emission wavelengths (typically 532 nm and 635 nm) and filter wavelengths (550-600 nm and 655-695 nm) to avoid background contamination. They are thus able to easily distinguish between two samples when one sample has been labeled with Cy3 and the other labeled with Cy5. They are also able to quantify the amount of labeling in either sample.

Cy Dye Alternatives

Alexa Fluor dyes, Atto Dyes, Dylight, IRIS Dyes, Seta dyes, SeTau dyes, SRfluor dyes and Square dyes dyes can be used interchangeably with Cy dyes in most biochemical applications.

Cy5 ozone Susceptibility

In 2003, researchers at Inpharmatics and Agilent reported in Analytical Chemistry that microarrays which used Cy5 were susceptible to intermittent data quality decrease caused by environmental ozone. Exposures to ozone levels above 5-10 ppb for 10–30 seconds were reported to decrease the reproducibility of Cy5 microarrays. Much higher levels of ozone (>100 ppb) were required to observe an effect in Cy3.

Applications

Nucleic Acid Labeling

In microarray experiments DNA or RNA is labeled with either Cy3 or Cy5 that has been synthesized to carry an N-hydroxysuccinimidyl ester (NHS-ester) reactive group. Since, NHS-esters react readily only with aliphatic amine groups, which nucleic acids lack, nucleotides have to be modified with aminoallyl groups. This is done through incorporating aminoallyl-modified nucleotides during synthesis reactions. A good ratio is a label every 60 bases such that the labels are not too close to each other, thus resulting in quenching effects.

Protein Labeling

For protein labeling, Cy3 and Cy5 dyes sometimes bear maleimide reactive groups instead. The maleimide functionality allows conjugation of the fluorescent dye to the sulfhydryl group of cysteine residues. Cysteines can be added and removed from the protein domain of interest via PCR mutagenesis.

Cy5, is sensitive to the electronic environment it resides in. Changes in the conformation of the protein it is attached to will produce an enhancement or quenching of the emission. The rate of this change can be measured to determine enzyme kinetic parameters. The dyes can be used for similar purposes in FRET experiments.

Cy3 and Cy5 are used in proteomics experiments so that samples from two sources can be mixed and run together thorough the separation process.

This eliminates variations due to differing experimental conditions that are inevitable if the samples were run separately. These variations make it extremely difficult, if not impossible, to use computers to automate the acquisition of the data after the separation is complete. Using these dyes makes the automation trivial.

Spectrophotometric Quantification of Dye Incorporation

The amount of each sample as well as the amount of incorporated dyes has to be quantified following nucleic acid or protein labeling and purification of the labeling reaction.

To determine the dye incorporation rate, the absorbance reading at the wavelength reported for maximum absorbance of the fluorescence dye is used. The corresponding extinction coefficient of the dye is used in the Lambert-Beer

Law to determine the dye concentration (c = A / (e * d)). Comparing these values with the DNA, RNA or protein concentration gives a dye incorporation rate (Frequency of incorporation (FOI)).

FOI = C(dye) / C(nucleic acid)

Example: Frequency of Incorporation (FOI) of Cy3 per 1000 bases: FOI(Cy3) = 58.5 * A550/A260

In multiple biological applications only minimal sample amounts can be afforded for these measurements (e.g. array CGH). Therefore, specialized photometers like the NanoPhotometer™ offer the possibility to determine sample concentrations as well as dye incorporation with submicroliter volumes (starting with 0.3 μl). In addition, due to the reduction of the optical pathlength samples are diluted automatically in comparison to standard cuvette measurements.

The respective virtual dilution factors are considered by the software of the instrument. Because the measurements are processed with undiluted samples, the reproducibility of the results is very high. And if desired, samples can be retrieved after the measurement for further processing.

Gene Chip Analysis

Microarray technology is a powerful tool for genomic analysis. It gives a global view of the genome in a single experiment. Data analysis of the microarray is a vital part of the experiment. Each microarray study comprises multiple microarrays, each giving tens of thousands of data points. Since the volume of data is growing exponentially as microarrays grow larger, the analysis becomes more challenging. In general the greater the volume of data, the more chances arise for erroneous results.

Handling such large volumes of data requires high end computational infrastructures and programs that can handle multiple data formats. There are already programs available for microarray data analysis on various platforms. However, due to rapid development, diversity in microarray technology, and different data formats, there is always the need for more comprehensive and complete microarray data analysis.

Data Analysis

Data analysis is critical, since any error introduced in the data analysis part will result in biologically insignificant results. In data analysis, the information from the raw data file is further processed to yield meaningful biological results. This part includes data normalization, flagging of the data, averaging the ratio for replicates, clustering of similarly expressed genes, etc. Data has to undergo normalization before further analysis. Normalization removes the non-biological variation between the samples. After the normalization, the ratio is calculated for each gene in the replicate. Based on the ratio, differentially regulated genes are determined. There are various statistical analyses which are also done for confidence analysis. Each replicate data is also examined for various experimental artifacts, bias by computing parameters related to intensity, background, flags, spot details, etc.

Replicates

It is important to note the necessity of replicates in conducting microarray experiments. Like any other quantitative measurement, repeated experiments provide the ability to conduct confidence analysis and identify differentially expressed genes at a given level of confidence. More replicates provide more confidence in determining differentially expressed genes. In practice, three to five replicates would be ideal.

Normalization

Normalization is required to standardize data and focus on biologically relevant changes. There are many sources of systematic variation in microarray experiments that affect the measured gene expression levels such as dye bias, heat and light sensitivity, efficiency of dye incorporation, differences in the labeled cDNA hybridization conditions, scanning conditions, and unequal quantities of starting RNA, etc.

Normalization is an important step in adjusting the data set for technical variation and removing relative abundance of gene expression profiles; this is the only point where 1- and 2-color data analyses vary. The normalization method depends on the data.

The basic idea behind all the normalization methods is that the expected mean intensity ratio between the two channels should be one. If the observed mean intensity ratio deviates from one, the data is mathematically processed in such a way that the final observed mean intensity ratio becomes one. With the mean intensity ratio adjusted to one, the distribution of the gene expression is centered so that genuine differentials can be identified.

Quality Control

Before analyzing data for biological variation, QC steps must be performed to determine whether the data is fit for statistical testing. Statistical tests are sensitive to the nature of the input data.

Filtering of Flagged Data

Filtering of bad intensity spots is an important process of quality control. For example, the scanner has a measurement limit below which intensity values cannot be trusted. Typically, the lowest intensity value of reliable data is 100–200 for Affymetrix data and 100–1000 for cDNA Microarray data. These cut-offs are likely to change as scanners become more precise. Values below the cut-off point are usually removed (filtered) from the data because they are likely to be artifacts.

Filtering of Noisy Replicates

Filtering of noisy replicates is a crucial part of quality control. Experimental replicates should have similar values. Replicates with noise should be eliminated before analysis; this can be done using the ANOVA statistical method.

Filtering of Non-significant Genes

Filtering of non-significant genes is done so that analysis can be done on selected genes. Non-significant genes are removed by specifying relative change in expression with respect to normal control. Values for over-expressed and under-expressed genes are defined as 2 and "2 respectively. As a result of filtering, few genes are retained. Those remaining genes are then subjected to statistical analysis.

Statistical Analysis

Statistical analysis plays a vital role in identifying genes that are expressed at statistically significant levels.

Clustering

Clustering is a data mining technique used to group genes having similar expression patterns. Hierarchical clustering, and k-mean clustering are widely used techniques in microarray analysis.

Hierarchical Clustering

Hierarchical clustering is a statistical method for finding relatively homogeneous clusters. Hierarchical clustering consists of two separate phases. Initially, a distance matrix containing all the pairwise distances between the genes is calculated. Pearson's correlation and Spearman's correlation are often used as dissimilarity estimates, but other methods, like Manhattan distance or Euclidean distance, can also be applied.

If the genes on a single chip are to be clustered, the Euclidean distance is the correct choice, since at least two chips are needed for calculation of any correlation measures. After calculation of the initial distance matrix, the hierarchical clustering algorithm either (A) joins iteratively the two closest clusters starting from single data points (agglomerative, bottom-up approach), or (B) partitions clusters iteratively starting from the complete set (divisive, top-down approach). After each step, a new distance matrix between the newly formed clusters and the other clusters is recalculated. Hierarchical cluster analysis methods include:

- Single linkage (minimum method, nearest neighbor)
- Complete linkage (maximum method, furthest neighbor)
- Average linkage (UPGMA).

K-means Clustering

K-means clustering is an algorithm for classifying and grouping genes based on pattern into K groups. Grouping is done by minimizing the sum of the squares of distances between the data and the corresponding cluster centroid. Thus the purpose of K-mean clustering is to classify data based on similar expression. (www.biostat.ucsf.edu).

Gene Ontology Studies

Gene ontology studies give biologically meaningful information about the gene including cellular location, molecular function, and biological function. This information is analyzed for differences in regulation in disease or drug treatment regimen, with respect to normal control.

Pathway Analysis

Pathway analysis gives specific information about the pathway being affected in disease conditions, with respect to normal control. Pathway analysis also allows identification of gene networks and how genes are regulated.

Tiling Array

Tiling Arrays are a subtype of microarray chips. Like traditional microarrays, they function by hybridizing labeled DNA or RNA target molecules to probes fixed onto a solid surface. Tiling arrays differ from traditional microarrays in the nature of the probes. Instead of probing for sequences of known or predicted genes which may be dispersed throughout the genome, tiling arrays probe intensively for sequences which are known to exist in a contiguous region of genome.

This is useful for characterizing regions of genome which are sequenced but with local functions that are largely unknown. Tiling arrays aid in transcriptome mapping as well as in discovering sites of DNA- protein interaction (ChIP-chip), of DNA methylation (MeDIP-chip), and of sensitivity to DNase (DNase Chip), in addition to other uses (eg array CGH). In addition to the advantage of being able to detect previously unidentified genes and regulatory sequences, improved quantification of transcription products is possible. Specific probes are present in millions of copies (as opposed to only several as in traditional arrays) within an array unit called a feature, with anywhere from 10,000 to more than 6,000,000 different features per array. Variable levels of mapping resolution are obtainable by adjusting the amount of sequence overlap between probes, or the amount of known base pairs between probe sequences, as well as the length of the probes themselves. For smaller genomes such as that of Arabidopsis, whole genomes can be examined. Tiling arrays are quickly becoming one of the most powerful tools in genome-wide investigations.

Synthesis and Manufacturers

There are two main ways of synthesizing tiling arrays; the first is photolithographic manufacturing and the second is mechanically spotting or printing. The first method involves in situ synthesis where probes, approximately 25bp, are built on the surface of the chip. These arrays can hold up to 6 million discrete features, each of which contains millions of copies of one probe. The other way of synthesizing tiling array chips is via mechanically printing probes onto the chip. This is done by using automated machines with pins that place the previously synthesized probes onto the surface. Due to the size restriction of the pins, these chips can hold up to nearly 400,000 features. Three manufacturers of tiling arrays are Affymetrix, NimbleGen and Agilent. Their products vary in probe length and spacing.

ChIP-on-chip

ChIP-on-chip (also known as ChIP-chip) is a technique that combines chromatin immunoprecipitation ("ChIP") with microarray technology ("chip"). Like regular ChIP, ChIP-on-chip is used to investigate interactions between proteins and DNA in vivo. Specifically, it allows the identification of the cistrome, sum of binding sites, for DNA-binding proteins on a genome-wide basis. Whole-genome analysis can be performed to determine the locations of binding sites for almost any protein of interest. As the name of the technique suggests, such proteins are generally those operating in the context of chromatin. The most prominent representatives of this class are transcription factors, replication-related proteins, like ORC, histones, their variants, and histone modifications. The goal of ChIP-on-chip is to localize protein binding sites that may help identify functional elements in the genome. For example, in the case of a transcription factor as a protein of interest, one can determine its transcription factor binding sites throughout the genome.

Other proteins allow the identification of promoter regions, enhancers, repressors and silencing elements, insulators, boundary elements, and sequences that control DNA replication. If histones are subject of interest, it is believed that the distribution of modifications and their localizations may offer new insights into the mechanisms of regulation.

One of the long-term goals ChIP-on-chip was designed for is to establish a catalogue of (selected) organisms that lists all protein-DNA interactions under various physiological conditions. This knowledge would ultimately help in the understanding of the machinery behind gene regulation, cell proliferation, and disease progression. Hence, ChIP-on-chip offers not only huge potential to complement our knowledge about the orchestration of the genome on the nucleotide level, but also on higher levels of information and regulation as it is propagated by research on epigenetics.

Technological Platforms

The technical platforms to conduct ChIP-on-chip experiments are DNA microarrays, or "chips". They can be classified and distinguished according to various characteristics:

- *Probe type:* DNA arrays can comprise either mechanically spotted cDNAs or PCR-products, mechanically spotted oligonucleotides, or oligonucleotides that are synthesized in situ. The early versions of microarrays were designed to detect RNAs from expressed genomic regions (open reading frames). Although such arrays are perfectly suited to study gene expression profiles, they have limited importance in ChIP experiments since most "interesting" proteins with respect to this technique bind in intergenic regions. Nowadays, even custom-made arrays can be designed and fine-tuned to match the requirements of an experiment. Also, any sequence of nucleotides can be synthesized to cover genic as well as intergenic regions.
- *Probe size:* Early version of cDNA arrays had a probe length of about 200bp. Latest array versions use oligos as short as 70- (Microarrays, Inc.) to 25-mers (Affymetrix). (Feb 2007)
- *Probe composition:* There are tiled and non-tiled DNA arrays. Non-tiled arrays use probes selected according to non-spatial criteria, i.e., the DNA sequences used as probes have no fixed distances in the genome. Tiled arrays, however, select a genomic region (or even a whole genome) and divide it into equal chunks. Such a region is called tiled path. The average distance between each pair of neighboring chunks (measured from the center of each chunk) gives the resolution of the tiled path. A path can be overlapping, end-to-end or spaced.
- *Array size:* The first microarrays used for ChIP-on-Chip contained about 13,000 spotted DNA segments representing all ORFs and intergenic regions from the yeast genome. Nowadays, Affymetrix offers whole-genome tiled yeast arrays with a resolution of 5bp (all in all 3.2 million probes). Tiled arrays for the human genome become more and more powerful, too. Just to name example, Affymetrix offers a set of seven arrays with about 90 million probes, spanning the complete non-repetitive part of the human genome with about 35bp spacing. (Feb 2007)

Besides the actual microarray, other hard- and software equipment is necessary to run ChIP-on-chip experiments. It is generally the case that one company's microarrays can not be analyzed by another company's processing hardware. Hence, buying an array requires also buying the associated workflow equipment. The most important elements are, among others, hybridization ovens, chip scanners, and software packages for subsequent numerical analysis of the raw data.

Workflow of a ChIP-on-chip Experiment

Starting with a biological question, a ChIP-on-chip experiment can be divided into three major steps: The first is to set up and design the experiment by selecting the appropriate array and probe type. Second, the actual experiment is performed in the wet-lab. Last, during the dry-lab portion of the cycle, gathered data are analyzed to either answer the initial question or lead to new questions so that the cycle can start again.

Wet-lab Portion of the Workflow

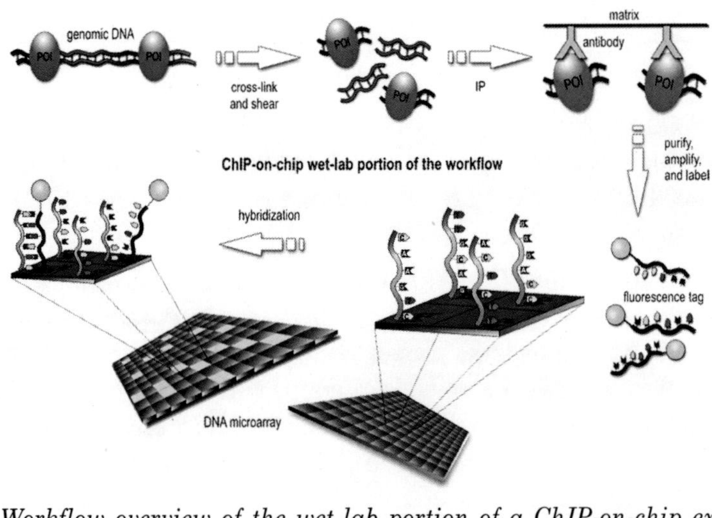

Figure: *Workflow overview of the wet-lab portion of a ChIP-on-chip experiment.*

- In the first step, the protein of interest (POI) is cross-linked with the DNA site it binds to in an in vivo environment. Usually this is done by a gentle formaldehyde fixation that is reversible with heat.
- Then, the cells are lysed and the DNA is sheared by sonication or using micrococcal nuclease. This results in double-stranded chunks of DNA fragments, normally 1 kb or less in length. Those that were cross-linked to the POI form a POI-DNA complex.
- In the next step, only these complexes are filtered out of the set of DNA fragments, using an antibody specific to the POI. The antibodies may

be attached to a solid surface, may have a magnetic bead, or some other physical property that allow distributing cross-linked complexes and unbound fragments. This procedure is essentially an immunoprecipitation (IP). There are two alternative ways to implement this filtering step:

- o immunoprecipitation of the tagged protein with an antibody against the tag (ex. FLAG, HA, c-myc)
- o affinity purification that does not require antibodies, such as the Tandem Affinity Purification (TAP)

- The cross-linking of POI-DNA complexes is reversed (usually by heating) and the DNA strands are purified. For the rest of the workflow, the POI is no longer necessary.
- After an amplification and denaturation step, the single-stranded DNA fragments are labeled with a fluorescent tag such as Cy5 or Alexa 647.
- Finally, the fragments are poured over the surface of the DNA microarray, which is spotted with short, single-stranded sequences that cover the genomic portion of interest. Whenever a labeled fragment "finds" a complementary fragment on the array, they will hybridize and form again a double-stranded DNA fragment.
- Quantitative and qualitative DNA and protein analysis with very small volumes of valuable samples are essential steps during the whole wet-lab workflow. Specialized NanoPhotometer offer the possibility of UV-visible spectrophotometry with sample volumes of 0.3 µl. Reduction of the optical pathlength, compared to standard measurements, effects an automatic dilution of the samples. Therefore, samples can be measured undiluted and if desired, retrieved after the measurement for further processing.

Dry-lab Portion of the Workflow

- After a sufficiently large time frame to allow hybridization, the array is illuminated with fluorescence light. Those probes on the array that are hybridized to one of the labeled fragments emit a light signal that is captured by a camera. This image contains all raw data for the remaining part of the workflow.
- This raw data, encoded as false-color image, needs to be converted to numerical values before the actual analysis can be done. The analysis and information extraction of the raw data often remains the most challenging part for ChIP-on-chip experiments. Problems arise throughout this portion of the workflow, ranging from the initial chip read-out, to suitable methods to subtract background noise, and finally to appropriate algorithms that normalize the data and make it available

for subsequent statistical analysis, which then hopefully lead to a better understanding of the biological question sought to answer. Furthermore, due to the different array platforms and missing standardization between them, data storage and exchange is a huge problem, too. Generally speaking, the data analysis can be divided into three major steps:

- During the first step, the captured fluorescence signals from the array are normalized, using control signals derived from the same or a second chip. Such control signals tell which probes on the array were hybridized correctly and which bound nonspecifically.
- In the second step, numerical and statistical tests are applied to control data and IP fraction data to identify POI-enriched regions along the genome. The following three methods are used widely: Median percentile rank, Single-array error, and Sliding-window. These methods generally differ in a way how low-intensity signals are handled, how much background noise is accepted, and which trait for the data is emphasized during the computation. In the recent past, the sliding-window approach seems to be favored and is often described as most powerful.
- In the third step, these regions are analyzed further. If, for example, the POI was a transcription factor, such regions would represent its binding sites. Subsequent analysis then may want to infer nucleotide motifs and other patterns to allow functional annotation of the genome.

Strengths and Weaknesses

Using tiled arrays, ChIP-on-chip allows for high resolution of genome-wide maps. These maps can determine the binding sites of many DNA-binding proteins like transcription factors and also chromatin modifications.

Although ChIP-on-chip can be a powerful technique in the area of genomics, it is very expensive. Most published studies using ChIP-on-chip repeat their experiments at least three times to ensure biologically meaningful maps. The cost of the DNA microarrays is often a limiting factor to whether a laboratory should proceed with a ChIP-on-chip experiment. Another limitation is the size of DNA fragments that can be achieved.

Most ChIP-on-chip protocols utilize sonication as a method of breaking up DNA into small pieces. However, sonication is limited to a minimal fragment size of 200 bp. For higher resolution maps, this limitation should be overcome to achieve smaller fragments, preferably to single nucleosome resolution. As mentioned previously, the statistical analysis of the huge amount of data generated from arrays is a challenge and normalization procedures should aim to minimize artifacts and determine what is really biologically significant. So far, application to mammalian genomes has been a major limitation, for example, due to a significant percentage of the genome that is occupied by repeats. However, as ChIP-on-chip technology advances, high resolution whole mammalian genome maps should become achievable.

Antibodies used for ChIP-on-chip can be an important limiting factor. ChIP-on-chip requires highly specific antibodies that must recognize its epitope in free solution and also under fixed conditions. If it is demonstrated to successfully immunoprecipitate cross-linked chromatin, it is termed "ChIP-grade". Companies that provide ChIP-grade antibodies include Abcam, Cell Signaling Technology, Santa Cruz, and Upstate.

To overcome the problem of specificity, the protein of interest can be fused to a tag like FLAG or HA that are recognized by antibodies. An alternative to ChIP-on-chip that does not require antibodies is DamID. Also available are antibodies against a specific histone modification like H3 tri methyl K4. As mentioned before, the combination of these antibodies and ChIP-on-chip has become extremely powerful in determining whole genome analysis of histone modification patterns and will contribute tremendously to our understanding of the histone code and epigenetics.A study demonstrating the non-specific nature of DNA binding proteins has been published in PLoS Biology. This indicates that alternate confirmation of functional relevancy is a necessary step in any ChIP-chip experiment.

History

A first ChIP-on-chip experiment was performed in 1999 to analyze the distribution of cohesin along budding yeast chromosome III. Although the genome was not completely represented, the protocol in this study remains equivalent as those used in later studies. The ChIP-on-chip technique using all of the ORFs of the genome (that nevertheless remains incomplete, missing intergenic regions) was then applied successfully in three papers published in 2000 and 2001.

The authors identified binding sites for individual transcription factors in the budding yeast Saccharomyces cerevisiae. In 2002, Richard Young's group determined the genome-wide positions of 106 transcription factors using a c-Myc tagging system in yeast. Other applications for ChIP-on-chip include DNA replication, recombination, and chromatin structure. Since then, ChIP-on-chip has become a powerful tool in determining genome-wide maps of histone modifications and many more transcription factors. ChIP-on-chip in mammalian systems has been difficult due to the large and repetitive genomes. Thus, many studies in mammalian cells have focused on select promoter regions that are predicted to bind transcription factors and have not analyzed the entire genome. However, whole mammalian genome arrays have recently become commercially available from companies like Nimblegen. In the future, as ChIP-on-chip arrays become more and more advanced, high resolution whole genome maps of DNA-binding proteins and chromatin components for mammals will be analyzed in more detail.

ChIP-chip

ChIP-chip is one of the most popular usages of tiling arrays. Chromatin immunoprecipitation is the technique whereby binding sites of proteins can be identified. A genome-wide variation of this is known as ChIP-on-chip. Proteins that bind to chromatin are cross-linked in vivo, usually via fixation with formaldehyde. The chromatin is then fragmented and exposed to antibodies specific to protein of interest. These complexes are then precipitated. The DNA is then isolated and purified. With traditional DNA microarrays, the immunoprecipitated DNA is hybridized to the chip, which contains probes, designed to cover regions representative of the genome. However, with tiling arrays, overlapping probes or probes in very close proximity can be used. This gives an unbiased analysis with high resolution. Besides these advantages, tiling arrays show high reproducibility, and with overlapping probes spanning large segments of the genome, tiling arrays can still interrogate protein binding sites, which harbor repeats. ChIP-chip experiments have been done to identify binding sites of transcription factors across the genome in yeast, drosophila and a few mammalian species.

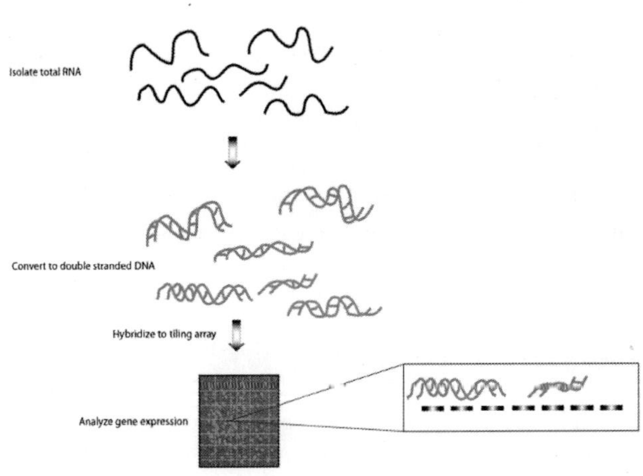

***Figure:** Overview of transcriptome mapping procedure.*

Transcriptome mapping

Another popular use of tiling arrays is in finding expressed genes. Traditional methods of gene prediction for annotation of genomic sequences have had several problems when used to map the transcriptome, such as not producing an accurate structure of the genes and also the missing of transcripts. The method of sequencing cDNA to find transcribed genes also runs into problems, like not being able to detect rare RNA molecules, RNA that are not polyadenylated, and very short RNA, and so would not detect genes that are only active in response to signals or specific to a time frame. Tiling arrays can solve these issues in mapping the transcriptome. Due to the high resolution and sensitivity, even small and rare molecules can be detected.

The overlapping nature of the probes also allows detection of non-polyadenylated RNA and can produce a more precise picture of gene structure. Earlier studies done on chromosome 21 and 22 showed the power of tiling arrays for identifying transcription units. The authors used 25-mer probes that were 35bp apart, spanning the entire chromosomes. Labeled targets were made from polyadenylated RNA. They found many more transcripts than predicted and 90% were outside of annotated exons. Another study done in Arabidopsis used high-density oligonucleotide arrays that cover the entire genome. More than 10 times more transcripts were found than predicted by ESTs and other prediction tools. Also found were novel transcripts in the centromeric regions where it was thought that no genes are actively expressed. Many noncoding and natural antisense RNA have also been identified using tiling arrays.

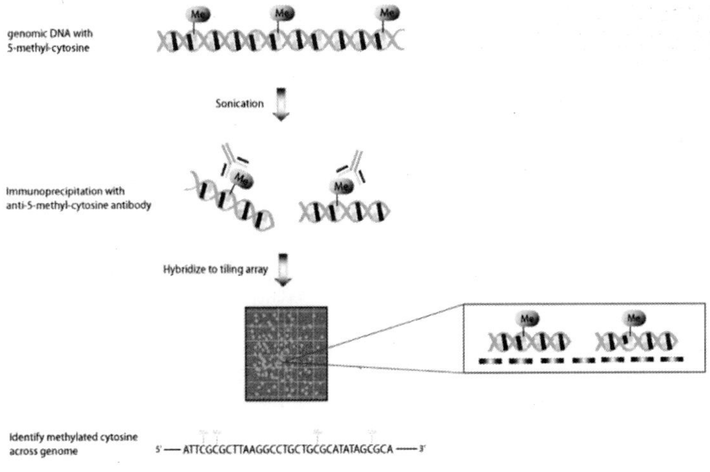

Figure: *Overview of MeDIP-chip procedure.*

MeDIP-chip

Methyl-DNA immunoprecipitation followed by tiling array allows DNA methylation mapping and measurement across the genome. DNA is methylated on cytosine in CG di-nucleotides in many places in the genome. This modification is one of the best-understood inherited epigenetic changes and is shown to affect gene expression. Mapping these sites can add to the knowledge of expressed genes and also epigenetic regulation on a genome-wide level. Studies have been done, utilizing tiling arrays, to generate high-resolution methylation maps for the Arabidopsis genome to generate the first "methylome".

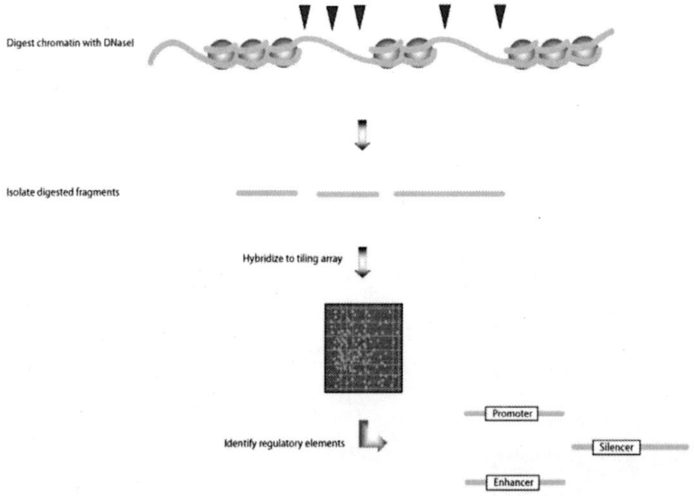

***Figure:** Overview of DNase-chip procedure.*

DNase-chip

DNase chip is an application of tiling arrays to identify hypersensitive sites, which are segments of open chromatin that are more readily cleaved by DNaseI. DNaseI cleaving produces larger fragments of around 1.2kb in size. These hypersensitive sites have been shown to be an accurate way of predicting regulatory elements such as promoter regions, enhancers and silencers. Historically, the method uses Southern blotting to find digested fragments; however, tiling arrays have been used in its place for applying the technique to a genome-wide scale.

Comparative Genomic Hybridization (CGH)

Array-based CGH is a technique often used in diagnostics to compare differences between types of DNA, such as normal cells vs. cancer cells. There are two types of tiling arrays commonly used for array CGH, which are the whole genome and fine tiled. The whole genome approach would be useful in identifying copy number variations with high resolution. On the other hand, the fine tiled array CGH would produce ultrahigh resolution to find other abnormalities such as breakpoints.

Procedure

There are several different methods for conducting a tiling arraying. One protocol for analyzing gene expression involves first isolating total RNA. This is then purified of rRNA molecules. The RNA is copied into double stranded DNA, which is subsequently amplified and in vitro transcribed to cRNA. The product is split into triplicates to produce dsDNA, which is then fragmented and labeled. Finally, the samples are hybridized to the tiling array chip. The signals from the chip are scanned and interpreted by computers.

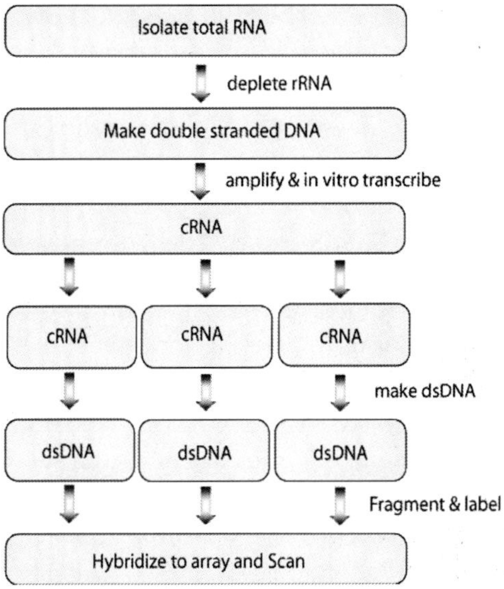

Figure: *Workflow of tiling array procedure.*

Various software and algorithms are available for data analysis and vary in benefits depending on the manufacturer of the array chip. For Affymetrix chips, the model-based analysis of tiling array (MAT) is the most effective peak-seeking algorithm. For NimbleGen chips, TAMAL is more suitable for locating binding sites. Alternative algorithms include MA2C and TileScope, which are less complicated to operate. The Joint binding deconvolution algorithm is commonly used for the Agilent chips. If sequence analysis of binding site or annotation of the genome is required then programs like MEME, Gibbs Motif Sampler, Cis-regulatory element annotation system and Galaxy are used.

Advantages and Disadvantages

The obvious advantages of tiling array are that they provide an unbiased tool to investigate protein binding, gene expression and gene structure on a genome-wide scope. They allow a new level of insight in studying the transcriptome and methylome.

However, there are also certain drawbacks. First and foremost is the issue of expense. Although the cost of purchasing tiling array kits have reduced in price in the last several years, at the moment, the price makes it impractical to actually use genome wide tiling arrays for larger genomes like mammalian. Another issue is to do with the ultra sensitive detection of this technology.

For looking at gene expression, an argument against a study in Arabidopsis sp., which found ten times more genes in the genome than traditional prediction tools, is that the results were confounded by "transcriptional noise". Furthermore there is an analysis challenge, as there is no clearly defined start or stop to your region of interest once identified on the array. Also, the arrays usually only give chromosome and position numbers, necessitating the need to sequence your region of interest (if required) as a separate step (although some modern arrays also give sequence information).

RNA-Seq

RNA-seq, also called "Whole Transcriptome Shotgun Sequencing" ("WTSS") and dubbed "a revolutionary tool for transcriptomics", refers to the use of high-throughput sequencing technologies to sequence cDNA in order to get information about a sample's RNA content, a technique that is quickly becoming invaluable in the study of diseases like cancer. Thanks to the deep coverage and base level resolution provided by next-generation sequencing instruments, RNA-seq provides researchers with efficient ways to measure transcriptome data experimentally, allowing them to get information such as how different alleles of a gene are expressed, detect post-transcriptional mutations or identify gene fusions.

Introduction

The introduction of next-generation sequencing or high-throughput sequencing technologies opened new doors into the field of DNA sequencing, however as understanding of these technologies becomes more widespread and new tools are developed, new innovative ways of applying these technologies are being created.

Given high-throughput sequencing technologies' low requirements of nucleotide sequence product, together with its deep coverage and base-scale resolution, its use has expanded to the field of transcriptomics. Transcriptomics is an area of research characterizing the RNA transcribed from a particular genome under investigation. Although transcriptomes are more dynamic than genomic DNA, these molecules provide direct access to gene regulation and protein information. Sequencing transcriptomes is not a new idea. Various methods have been developed previously to directly determine cDNA sequences based mostly around traditional (and more expensive) Sanger sequencing, while others include methodologies such as Serial analysis of gene expression (SAGE), cap analysis gene expression (CAGE) and massively parallel signature sequencing (MPSS).

Transcriptome Sequencing (RNA-seq) can be done with a variety of platforms to test a high amount of ideas and hypotheses. For example, using the Illumina (company) Genome Analyzer platform, recent applications include sequencing mammalian transcriptomes, ABI Solid Sequencing to profile stem cell transcriptomes or Life Science's 454 Sequencing to discover SNPs in maize. Even though each platform has its technical differences, the information gathered from each is of the same nature.

Methods

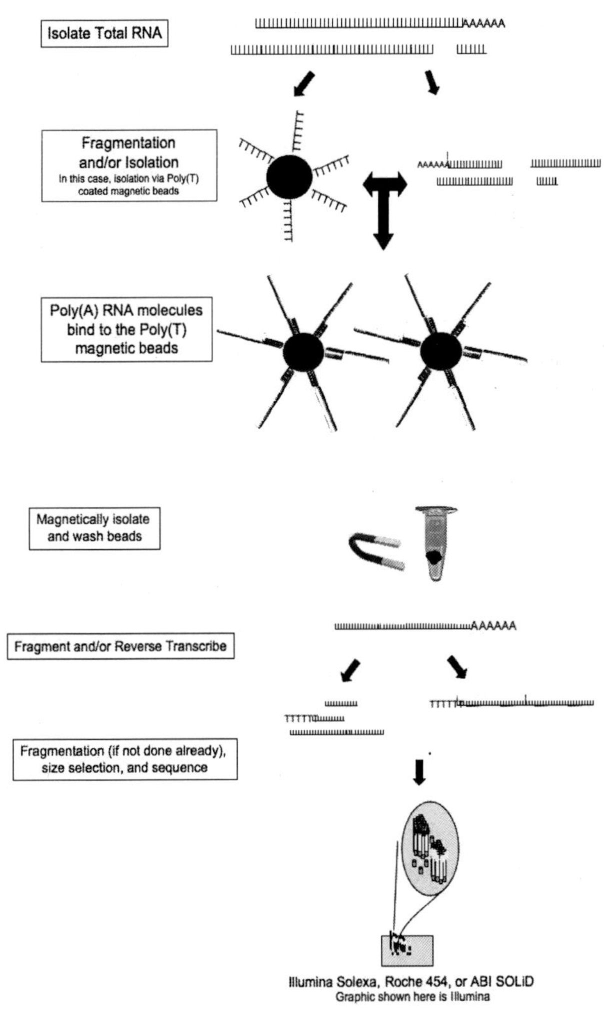

RNA Poly(A) Library

Creation of a sequence library can change from platform to platform in high throughput sequencing, where each has several kits designed to build different types of libraries and adapting the resulting sequences to the specific requirements of their instruments.

However, due to the nature of the template being analyzed, there are commonalities within each technology. Frequently, in mRNA analysis the 3' polyadenylated (poly(A)) tail is targeted in order to ensure that coding RNA is separated from noncoding RNA. This can be accomplished simply with poly (T) oligos covalently attached to a given substrate. Presently many studies utilize magnetic beads for this step (;). The Protocol Online website provides a list of several protocols relating to mRNA isolation.

Studies including portions of the transcriptome outside poly(A) RNAs have shown that when using poly(T) magnetic beads, the flow-through RNA (non-poly(A) RNA) can yield important noncoding RNA gene discovery which would have otherwise gone unnoticed.

Also, since ribosomal RNA represents over 90% of the RNA within a given cell, studies have shown that its removal via probe hybridization increases the capacity to retrieve data from the remaining portion of the transcriptome.

The next step is reverse transcription. Due to the 5' bias of randomly primed-reverse transcription as well as secondary structures influencing primer binding sites, hydrolysis of RNA into 200-300 nucleotides prior to reverse transcription reduces both problems simultaneously.

However, there are trade-offs with this method where although the overall body of the transcripts are efficiently converted to DNA, the 5' and 3' ends are less so. Depending on the aim of the study, researchers may choose to apply or ignore this step.

Once the cDNA is synthesized it can be further fragmented to reach the desired fragment length as specified in table 1. The template is now ready to be prepared for the desired sequencing method.

Polyadenylation

Polyadenylation is the addition of a poly(A) tail to an RNA molecule. The poly(A) tail consists of multiple adenosine monophosphates; in other words, it is a stretch of RNA that has only adenine bases. In eukaryotes, polyadenylation is part of the process that produces mature messenger RNA (mRNA) for translation. It, therefore, forms part of the larger process of gene expression.

The process of polyadenylation begins as the transcription of a gene finishes. The 3'-most segment of the newly-made RNA is first cleaved off by a set of proteins; these proteins then synthesize the poly(A) tail at the RNA's 3' end. In some genes, these proteins may add a poly(A) tail at any one of several possible sites. Therefore, polyadenylation can produce more than one transcript from a single gene, similar to alternative splicing.

The poly(A) tail is important for the nuclear export, translation, and stability of mRNA. The tail is shortened over time, and, when it is short enough, the mRNA is enzymatically degraded. However, in a few cell types, mRNAs with short poly(A) tails are stored for later activation by re-polyadenylation in the cytosol.

In contrast, when polyadenylation occurs in bacteria, it promotes RNA degradation. This is also sometimes the case for eukaryotic non-coding RNAs. The wide distribution of polyadenylation among living organisms indicates that this process evolved early in the history of life on Earth.

Primer on RNA

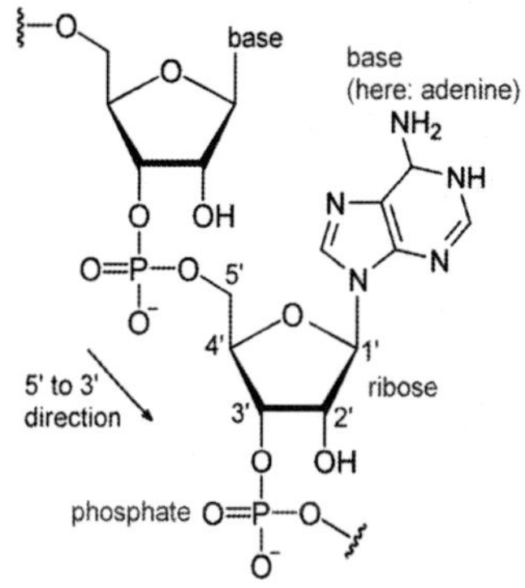

Figure: *Chemical structure of RNA. The sequence of bases differs between RNA molecules.*

RNAs are a type of large biological molecules, whose individual building blocks are called nucleotides. The name poly(A) tail (for polyadenylic acid tail) reflects the way RNA nucleotides are abbreviated, with a letter for the base the nucleotide contains (A for adenine, C for cytosine, G for guanine and U for uracil). RNAs are produced (transcribed) from a DNA template. By convention, RNA sequences are written in a 5' to 3' direction.

The 5' end is the part of the RNA molecule that is transcribed first, and the 3' end is transcribed last. The 3' end is also where the poly(A) tail is found on polyadenylated RNAs. Messenger RNA (mRNA) is RNA that has a coding region that acts as a template for protein synthesis (translation). The rest of the mRNA, the untranslated regions, tune how active the mRNA is. There are also many RNAs that are not translated, called non-coding RNAs. Like the untranslated regions, many of these non-coding RNAs have regulatory roles.

Nuclear Polyadenylation

Function

In nuclear polyadenylation, a poly(A) tail is added to an RNA at the end of transcription. On mRNAs, the poly(A) tail protects the mRNA molecule from enzymatic degradation in the cytoplasm and aids in transcription termination, export of the mRNA from the nucleus, and translation. Almost all eukaryotic mRNAs are polyadenylated, with the exception of animal replication-dependent histone mRNAs. These are the only mRNAs in eukaryotes that lack a poly(A) tail, ending instead in a stem-loop structure followed by a purine-rich sequence, termed histone downstream element, that directs where the RNA is cut so that the 3' end of the histone mRNA is formed.

Many eukaryotic non-coding RNAs are always polyadenylated at the end of transcription. There are small RNAs where the poly(A) tail is seen only in intermediary forms and not in the mature RNA as the ends are removed during processing, the notable ones being microRNAs. But, for many long noncoding RNAs – a seemingly large group of regulatory RNAs that, for example, includes the RNA Xist, which mediates X chromosome inactivation – a poly(A) tail is part of the mature RNA.

Mechanism

The polyadenylation machinery in the nucleus of eukaryotes works on products of RNA polymerase II, such as precursor mRNA. Here, a multi-protein complex (see components on the right) cleaves the 3'-most part of a newly produced RNA and polyadenylates the end produced by this cleavage. The cleavage is catalysed by the enzyme CPSF and occurs 10–30 nucleotides downstream of its binding site. This site is often the sequence AAUAAA on the RNA, but variants of it that bind more weakly to CPSF exist. Two other proteins add specificity to the binding to an RNA: CstF and CFI. CstF binds to a GU-rich region further downstream of CPSF's site. CFI recognises a third site on the RNA (a set of UGUAA sequences in mammals) and can recruit CPSF even if the AAUAAA sequence is missing. The polyadenylation signal – the sequence motif recognised by the RNA cleavage complex – varies between groups of eukaryotes. Most human polyadenylation sites contain the AAUAAA sequence, but this sequence is less common in plants and fungi.

The RNA is cleaved right after transcription, as CstF also binds to RNA polymerase II. Cleavage also involves the protein CFII, though it is unknown how. The cleavage site associated with a polyadenylation signal can vary up to some 50 nucleotides. When the RNA is cleaved, polyadenylation starts, catalysed by polyadenylate polymerase. Polyadenylate polymerase builds the poly(A) tail by adding adenosine monophosphate units from adenosine triphosphate to the RNA, cleaving off pyrophosphate. Another protein, PAB2, binds to the new, short poly(A) tail and increases the affinity of polyadenylate polymerase for the RNA. When the poly(A) tail is approximately 250 nucleotides long the enzyme can no longer bind to CPSF and polyadenylation stops, thus determining the length of the poly(A) tail. CPSF is in contact with RNA polymerase II, allowing it to tell the polymerase to terminate transcription. The polyadenylation machinery is also physically linked to the spliceosome, a complex that removes introns from RNAs.

Downstream Effects

The poly(A) tail acts as the binding site for poly(A)-binding protein. Poly(A)-binding protein promotes export from the nucleus and translation, and inhibits degradation. This protein binds to the poly(A) tail prior to mRNA export from the nucleus and in yeast also recruits poly(A) nuclease, an enzyme that shortens the poly(A) tail and allows the export of the mRNA. Poly(A)-binding protein is exported to the cytoplasm with the RNA. mRNAs that are not exported are degraded by the exosome. Poly(A)-binding protein also can bind to, and thus recruit, several proteins that affect translation, one of these is initiation factor-4G, which in turn recruits the 40S ribosomal subunit. However, a poly(A) tail is not required for the translation of all mRNAs.

Deadenylation

In eukaryotic somatic cells, the poly(A) tail of most mRNAs in the cytoplasm gradually get shorter, and mRNAs with shorter poly(A) tail are translated less and degraded sooner. However, it can take many hours before an mRNA is degraded. This deadenylation and degradation process can be accelerated by microRNAs complementary to the 3' untranslated region of an mRNA. In immature egg cells, mRNAs with shortened poly(A) tails are not degraded, but are instead stored without being translated. They are then activated by cytoplasmic polyadenylation after fertilisation, during egg activation.

In animals, poly(A) ribonuclease (PARN) can bind to the 5' cap and remove nucleotides from the poly(A) tail. The level of access to the 5' cap and poly(A) tail is important in controlling how soon the mRNA is degraded. PARN deadenylates less if the RNA is bound by the initiation factors 4E (at the 5' cap) and 4G (at the poly(A) tail), which is why translation reduces deadenylation. The rate of deadenylation may also be regulated by RNA-binding proteins. Once the poly(A) tail is removed, the decapping complex removes the 5' cap, leading to a degradation of the RNA. Several other enzymes that seem to be involved in deadenylation have been identified in yeast.

Alternative Polyadenylation

Many protein-coding genes have more than one polyadenylation site, so a gene can code for several mRNAs that differ in their 3' end. Since alternative polyadenylation changes the length of the 3' untranslated region, it can change which binding sites for microRNAs the 3' untranslated region contains. MicroRNAs tend to repress translation and promote degradation of the mRNAs they bind to, although there are examples of microRNAs that stabilise transcripts. Alternative polyadenylation can also shorten the coding region, thus making the mRNA code for a different protcin, but this is much less common than just shortening the 3' untranslated region.

The choice of poly(A) site can be influenced by extracellular stimuli and depends on the expression of the proteins that take part in polyadenylation. For example, the expression of CstF-64, a subunit of cleavage stimulatory factor (CstF), increases in macrophages in response to lipopolysaccharides (a group of bacterial compounds that trigger an immune response). This results in the selection of weak poly(A) sites and thus shorter transcripts. This removes regulatory elements in the 3' untranslated regions of mRNAs for defense-related products like lysozyme and TNF-á. These mRNAs then have longer half-lives and produce more of these proteins. RNA-binding proteins other than those in the polyadenylation machinery can also affect whether a polyadenyation site is used,as can DNA methylation near the polyadenylation signal.

Cytoplasmic Polyadenylation

There is polyadenylation in the cytosol of some animal cell types, namely in the germ line, during early embryogenesis and in post-synaptic sites of nerve cells. This lengthens the poly(A) tail of an mRNA with a shortened poly(A) tail, so that the mRNA will be translated. These shortened poly(A) tails are often less than 20 nucleotides, and are lengthened to around 80–150 nucleotides. In the early mouse embryo, cytoplasmic polyadenylation of maternal RNAs from the egg cell allows the cell to survive and grow even though transcription does not start until the middle of the 2-cell stage (4-cell stage in human). In the brain, cytoplasmic polyadenylation is active during learning and could play a role in long-term potentiation, which is the strengthening of the signal transmission from a nerve cell to another in response to nerve impulses and is important for learning and memory formation.

Cytoplasmic polyadenylation requires the RNA-binding proteins CPSF and CPEB, and can involve other RNA-binding proteins like Pumilio. Depending on the cell type, the polymerase can be the same type of polyadenylate polymerase (PAP) that is used in the nuclear process, or the cytoplasmic polymerase GLD-2.

Tagging for Degradation in Eukaryotes

For many non-coding RNAs, including tRNA, rRNA, snRNA, and snoRNA, polyadenylation is a way of marking the RNA for degradation, in at least yeast. This polyadenylation is done in the nucleus by the TRAMP complex, which adds a tail that is around 40 nucleotides long to the 3' end. The RNA is then degraded by the exosome. Poly(A) tails have also been found on human rRNA fragments, both the form of homopolymeric (A only) and heterpolymeric (mostly A) tails.

In Prokaryotes and Organelles

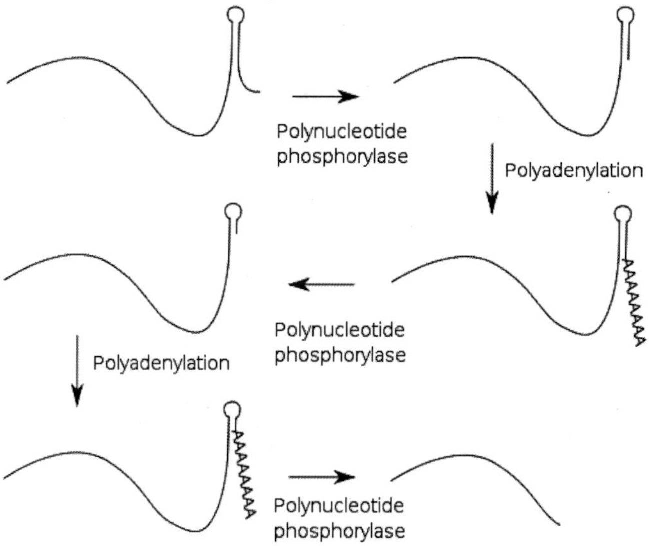

Figure: *Polyadenylation in bacteria helps polynucleotide phosphorylase degrade past secondary structure.*

In many bacteria, both mRNAs and non-coding RNAs can be polyadenylated. This poly(A) tail promotes degradation by the degradosome, which contains two RNA-degrading enzymes: polynucleotide phosphorylase and RNase E. Polynucleotide phosphorylase binds to the 3' end of RNAs and the 3' extension provided by the poly(A) tail allows it to bind to the RNAs whose secondary structure would otherwise block the 3' end. Successive rounds of polyadenylation and degradation of the 3' end by polynucleotide phosphorylase allows the degradosome to overcome these secondary structures. The poly(A) tail can also recruit RNases that cut the RNA in two. These bacterial poly(A) tails are about 30 nucleotides long.

In as different groups as animals and trypanosomes, the mitochondria contain both stabilising and destabilising poly(A) tails. Destabilising polyadenylation targets both mRNA and noncoding RNAs. The poly(A) tails are 43 nucleotides long on average. The stabilising ones start at the stop codon, and without them the stop codon (UAA) is not complete as the genome only encodes the U or UA part. Plant mitochondria have only destabilising polyadenylation, and yeast mitochondria have no polyadenylation at all.

While many bacteria and mitochondria have polyadenylate polymerases, they also have another type of polyadenylation, performed by polynucleotide phosphorylase itself. This enzyme is found in bacteria, mitochondria, plastids and as a constituent of the archeal exosome (in those archaea that have an exosome). It can synthesise a 3' extension where the vast majority of the bases are adenines. Like in bacteria, polyadenylation by polynucleotide phosphorylase promotes degradation of the RNA in plastids and likely also archaea.

Evolution

Although polyadenylation is seen in almost all organisms, it is not universal. However, the wide distribution of this modification and the fact that it is present in organisms from all three domains of life implies that the last universal common ancestor of all living organisms, it is presumed, had some form of polyadenylation system.

A few organisms do not polyadenylate mRNA, which implies that they have lost their polyadenylation machineries during evolution. Although no examples of eukaryotes that lack polyadenylation are known, mRNAs from the bacterium Mycoplasma gallisepticum and the salt-tolerant archaean Haloferax volcanii lack this modification.

The most ancient polyadenylating enzyme is polynucleotide phosphorylase. This enzyme is part of both the bacterial degradosome and the archaeal exosome, two closely related complexes that recycle RNA into nucleotides. This enzyme degrades RNA by attacking the bond between the 3'-most nucleotides with a phosphate, breaking off a diphosphate nucleotide.

This reaction is reversible, and so the enzyme can also extend RNA with more nucleotides. The heteropolymeric tail added by polynucleotide phosphorylase is very rich in adenine. The choice of adenine is most likely the result of higher ADP concentrations than other nucleotides as a result of using ATP as an energy currency, making it more likely to be incorporated in this tail in early lifeforms. It has been suggested that the involvement of adenine-rich tails in RNA degradation prompted the later evolution of polyadenylate polymerases (the enzymes that produce poly(A) tails with no other nucleotides in them).

Polyadenylate polymerases are not as ancient. They have separately evolved in both bacteria and eukaryotes from CCA-adding enzyme, which is the enzyme that completes the 3' ends of tRNAs. Its catalytic domain is homologous to that of other polymerases. It is presumed that the horizontal transfer of bacterial CCA-adding enzyme to eukaryotes allowed the archaeal-like CCA-adding enzyme to switch function to a poly(A) polymerase. Some lineages, like archaea and cyanobacteria, never evolved a polyadenylate polymerase.

History

Polyadenylation was first identified in 1960 as an enzymatic activity in extracts made from cell nuclei that could polymerise ATP, but not ADP, into polyadenine. Although identified in many types of cells, this activity had no known function until 1971, when poly(A) sequences were found in mRNAs. The only function of these sequences was thought at first to be protection of the 3' end of the RNA from nucleases, but later the specific roles of polyadenylation in nuclear export and translation were identified. The polymerases responsible for polyadenylation were first purified and characterized in the 1960s and 1970s, but the large number of accessory proteins that control this process were discovered only in the early 1990s.

Direct RNA Sequencing (DRS™) from Helicos' Company

Methods for in-depth characterization of transcriptomes and quantification of transcript levels have emerged as valuable tools for understanding cellular physiology and human disease biology, and have begun to be utilized in various clinical diagnostic applications. Current methods, however, typically require RNA to be converted to complementary DNA (cDNA) via reverse transcription prior to measurements. This step has been shown to introduce biases and artifacts that may interfere with both the proper characterization and quantification of transcripts. To enable virtually unbiased view and quantification of transcriptomes, Helicos has developed and commercialized the single molecule Direct RNA Sequencing (DRSTM) technology. DRSTM is the first and currently the only technology that can sequence RNA molecules directly in a massively-parallel manner without RNA conversion to cDNA or other biasing sample manipulations such as ligation and amplification.

Transcriptome Alignment

Due to the small size of the short reads (for Illumina Genome Analyzer this can be around 36 bases) de novo assembly may be difficult (though some software does exist:

Velvet (algorithm)), as there cannot be large overlaps between each read needed to easily reconstruct the original sequences, and the deep coverage makes the computing power to track all the possible alignments prohibitive. This can be somewhat overcome by having larger sequences obtained from the same sample using other techniques as Sanger sequencing, and using larger reads as a "skeleton" or a "template" to help assemble reads in difficult regions (e.g. regions with repetitive sequences).

The recommended approach is that of aligning the millions of reads to a "reference genome". There are many tools available for aligning genomic reads to a reference genome (sequence alignment tools), however, special attention is needed when alignment of a transcriptome to a genome, mainly when dealing with genes having intronic regions. As discussed above, the sequence libraries are created extracting mRNA using its poly(A) tail, which is added to the mRNA molecule post-transcriptionally and thus splicing has taken place. Therefore, the created library and the short reads obtained cannot come from intronic sequences and thus, when trying to align these short reads to a reference genome, only short reads aligning entirely inside exonic regions will be matched while short reads from exon-exon junction regions will not.

A possible method to work around this is to try to align the unaligned short reads using a proxy genome generated with known exonic sequences. This need not cover whole exons, only enough so that the short reads can match on both sides of the exon-exon junction with minimum overlap.

The use of paired-end sequencing has been mentioned as a good solution to alignment problems, as besides giving longer length reads, it allows obtaining information in respect to the strand. Several software packages exist for short read alignment, and recently specialized algorithms for transcriptome alignment have been developed, e.g. TopHat and Cufflinks.

Analysis

Gene Expression

The characterization of gene expression in cells via measurement of mRNA levels has long been of interest to researchers. Even though it has been shown that due to other post transcriptional gene regulation events (such as RNA interference) there is not a strong correlation between the abundance of mRNA and the related proteins, measuring mRNA concentration levels is still a useful tool in determining how the transcriptional machinery of the cell is affected in the presence of external signals (e.g. drug treatment), or how cells differ between a healthy state and a diseased state.

Microarray Approach

Prior to RNA-seq, DNA microarrays were unchallenged as the experiment of choice for transcriptome analysis. Although many experiments are still using microarrays to generate exciting results where the amount of time to retrieve results for a given sample is shorter, intrinsic experimental limitations of microarrays seem to make RNA-seq the method of choice. One important limitation is a prerequisite for sequence information in order to detect and ultimately evaluate transcripts.

As research in the field of RNA-seq is growing steadily with promising and consistent results, one must now consider, "Is this the beginning of the end for microarrays?". Still, for many applications, microarrays are the method of choice. Not only are they 10-100 times cheaper when compared at the same resolution of accuracy (less than $100 per array for high-throughput applications), but RNA-seq protocols still suffer from unknown biases such as those implied by the required ligation steps.

Another limitation for quantitative expression profiling is that, by design, high abundance transcripts (such as from housekeeping genes) are responsible for the majority of the sequencing data.

In a typical sample, 5% of the genes give rise to 75% of the reads sequenced. As a consequence, it is hard to measure the abundances of the remaining genes reliably, and the majority of transcript measurements is very noisy.

Coverage as Measure of Expression

Expression can be deduced via RNA-seq to the extent at which a sequence is retrieved. Transcriptome studies in yeast show that in this experimental setting, a fourfold coverage is required for amplicons to be classified and characterized as an expressed gene.

When the transcriptome is fragmented prior to cDNA synthesis, the number of reads corresponding to the particular exon normalized by its length in vivo yields gene expression levels which correlate with those obtained through qPCR. Single nucleotide polymorphism

A single-nucleotide polymorphism (SNP, pronounced snip) is a DNA sequence variation occurring when a single nucleotide — A, T, C or G — in the genome (or other shared sequence) differs between members of a biological species or paired chromosomes in an individual.

For example, two sequenced DNA fragments from different individuals, AAGCCTA to AAGCTTA, contain a difference in a single nucleotide. In this case we say that there are two alleles: C and T. Almost all common SNPs have only two alleles. SNPs can occur in both coding and non-coding regions of genome.

Within a population, SNPs can be assigned a minor allele frequency — the lowest allele frequency at a locus that is observed in a particular population. This is simply the lesser of the two allele frequencies for single-nucleotide polymorphisms. There are variations between human populations, so a SNP allele that is common in one geographical or ethnic group may be much rarer in another.

These genetic variations between the individuals (particularly in the non-coding parts of genome) are exploited in DNA fingerprinting, which is used in criminology. Also, these genetic variations underlie differences in our susceptibility to, or protection from all kinds of diseases.

The severity of illness and the way our body responds to treatments are also manifestations of genetic variations. For example, a single base difference in the Apolipoprotein E is associated with Alzheimer's disease.

Types

Types of SNPs:

- Non-coding region
- Coding region
    - Synonymous
    - Nonsynonymous
- Missense
- Nonsense

Single-nucleotide polymorphisms may fall within coding sequences of genes, non-coding regions of genes, or in the intergenic regions (regions between genes). SNPs within a coding sequence do not necessarily change the amino acid sequence of the protein that is produced, due to degeneracy of the genetic code. A SNP in which both alleles produce the same polypeptide sequence is called a synonymous polymorphism (sometimes called a silent mutation).

If a different polypeptide sequence is produced the polymorphism is a replacement polymorphism. A replacement polymorphism change may be either missense, which results in a different amino acid, or nonsense, which results in a premature stop codon. Over half of all known disease mutations come from replacement polymorphisms.

SNPs that are not in protein-coding regions may still affect gene splicing, transcription factor binding, or the sequence of non-coding RNA. Gene expression affected by this type of SNP is referred to as an eSNP (expression SNP) and may be upstream or downstream from the gene.

Use and Importance

Variations in the DNA sequences of humans can affect how humans develop diseases and respond to pathogens, chemicals, drugs, vaccines, and other agents. SNPs are also thought to be key enablers in realizing the concept of personalized medicine.

However, their greatest importance in biomedical research is for comparing regions of the genome between cohorts (such as with matched cohorts with and without a disease) in genome-wide association studies.

They are usually biallelic and thus easily assayed. SNPs do not usually function individually, rather, they work in coordination with other SNPs to manifest a disease condition as has been seen in osteoporosis.

Single Nucleotide Variation Discovery

Transcriptome single nucleotide variation has been analyzed in maize on the Roche 454 sequencing platform. Directly from the transcriptome analysis, around 7000 single nucleotide polymorphisms (SNPs) were recognized. Following Sanger sequence validation, the researchers were able to conservatively obtain almost 5000 valid SNPs covering more than 2400 maize genes. This impressive transcriptome analysis is currently being applied to cancer research and microbiology which could quite possibly lead to new forms of medicine.

Coverage/Depth

Coverage/depth can affect the mutations seen and given that everything is expression-centric, an allele might not be detected either because it is not in the genome, or because it is not being expressed. At the same time, RNA-seq can yield additional information rather than just the existence of a heterozygous gene as it can also help in estimating the expression of each allele. In association studies, genotypes are associated to disease and expression levels can also be associated with disease.

Using RNA-seq, we can measure the relationship between these two associated variables, that is, in what relation are each of the alleles being expressed. The depth of sequencing required for specific applications can be extrapolated from a pilot experiment.

Germline vs Expressed Alleles

The only way to be absolutely sure of the individual's mutations is to compare the transcriptome sequences to the germline DNA sequence. This enables the distinction of homozygous genes versus skewed expression of one of the alleles and it can also provide information about genes that were not expressed in the transcriptomic experiment.

Post-transcriptional SNVs

Having the matching genomic and transcriptomic sequences of an individual can also help in detecting post-transcriptional edits, where, if the individual is homozygous for a gene, but the gene's transcript has a different allele, then a post-transcriptional modification event is determined.

mRNA centric single nucleotide variants (SNVs) are generally not considered as a representative source of functional variation in cells, mainly due to the fact that these mutations disappear with the mRNA molecule, however the fact that efficient DNA correction mechanisms do not apply to RNA molecules can cause them to appear more often. This has been proposed as the source of certain prion diseases, also known as TSE or transmissible spongiform encephalopathies.

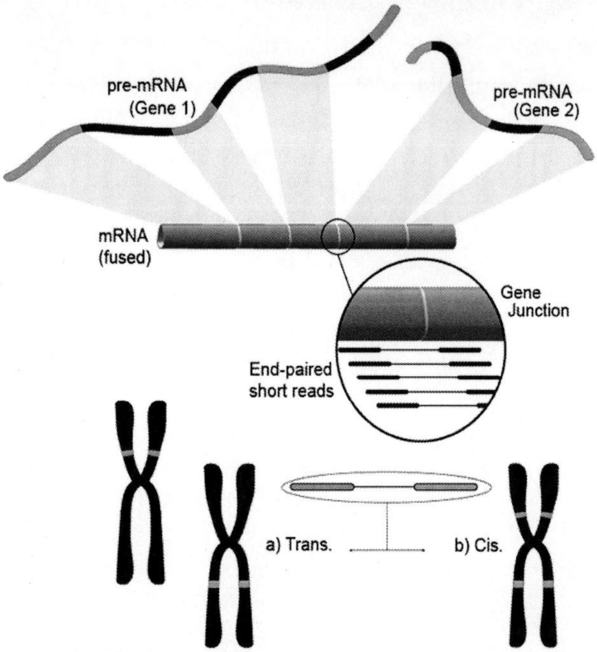

Figure: *RNA-seq mapping of short reads over exon-exon junctions, depending on where each end maps to, it could be defined a Trans or a Cis event.*

Fusion Gene

A fusion gene is a hybrid gene formed from two previously separate genes. It can occur as the result of a translocation, interstitial deletion, or chromosomal inversion. Often, fusion genes are oncogenes; examples include BCR-ABL, TEL-AML1 (ALL with t(12 ; 21)), AML1-ETO (M2 AML with t(8 ; 21)) and TMPRSS2-ERG with an interstitial deletion on chromosome 21, often occurring in prostate cancer. Most fusion genes are found from hematological cancers, sarcomas and prostate cancer.

Oncogenic fusion genes may lead to a gene product with a new or different function from the two fusion partners. Alternatively, a proto-oncogene is fused to a strong promoter, and thereby the oncogenic function is set to function by an upregulation caused by the strong promoter of the upstream fusion partner. The latter is common in lymphomas, where oncogenes are juxtaposed to the promoters of the immunoglobulin genes. Oncogenic fusion transcripts may also be caused by trans-splicing or read-through events.

Presence of certain chromosomal aberrations and their resulting fusion genes is commonly used within cancer diagnostics in order to set a precise diagnosis. In routine diagnostic laboratories the following methods are most commonly used: chromosome banding analysis, fluorescence in situ hybridisation, and RT-PCR. These methods all have their distinct shortcomings due to the very complex nature of cancer genomes. Recent developments such as high-throughput sequencing and custom-designed DNA microarrays bear promise of introduction of more efficient methods.

Biologists may also deliberately create fusion genes for research purposes. For example, by creating a fusion gene of a protein of interest and green fluorescent protein, the protein of interest may be observed in cells or tissue using fluorescence microscopy. The protein synthesized when a fusion gene is expressed is called a fusion protein.

Fusion Gene Detection

Caused by different structural modifications in the genome, fusion genes have gained attention because of their relationship with cancer. The ability of RNA-seq to analyze a sample's whole transcriptome in an unbiased fashion makes it an attractive tool to find these kinds of common events in cancer.

The idea follows from the process of aligning the short transcriptomic reads to a reference genome. Most of the short reads will fall within one complete exon, and a smaller but still large set would be expected to map to known exon-exon junctions. The remaining unmapped short reads would then be further analyzed to determine whether they match an exon-exon junction where the exons come from different genes.

This would be evidence of a possible fusion event, however, because of the length of the reads, this could prove to be very noisy. An alternative approach is to use pair-end reads, when a potentially large number of paired reads would map each end to a different exon, giving better coverage of these events.

Nonetheless, the end result consists of multiple and potentially novel combinations of genes providing an ideal starting point for further validation.

Some Considerations

The information gathered when sequencing a sample's transcriptome in this way has many of the same limitations as other RNA expression analysis pipelines. Mainly, the information gathered is:

a) Tissue specific: Gene expression is not uniform throughout an organism's cells, it is strongly dependent on the tissue type being measured;

b) Time dependent: During a cell's lifetime and context, its gene expression levels change.

Because of this, care must be taken when drawing conclusions from the sequencing experiment, as some of the information gathered might not be representative of the individual itself. An example of this would be during SNV discovery as the mutations discovered are more precisely the mutations being expressed, this is: observing a homozygote location to a non-reference allele in an organism does not necessarily mean that this is the individual's genotype, it could just mean that the gene copy with the reference allele is not being expressed in that tissue and/or at the time snapshot the sample was acquired.

Gene Family

A gene family is a set of several similar genes, formed by duplication of a single original gene, and generally with similar biochemical functions. One such family are the genes for human haemoglobin subunits; the ten genes are in two clusters on different chromosomes, called the á-globin and â-globin loci.

Genes are categorized into families based on shared nucleotide or protein sequences. Phylogenetic techniques can be used as a more rigorous test. The positions of exons within the coding sequence can be used to infer common ancestry. Knowing the sequence of the protein encoded by a gene can allow researchers to apply methods that find similarities among protein sequences that provide more information than similarities or differences among DNA sequences. Furthermore, knowledge of the protein's secondary structure gives further information about ancestry, since the organization of secondary structural elements presumably would be conserved even if the amino acid sequence changes considerably. These methods often rely upon predictions based upon the DNA sequence.

If the genes of a gene family encode proteins, the term protein family is often used in an analogous manner to gene family.

The expansion or contraction of gene families along a specific lineage can be due to chance, or can be the result of natural selection. To distinguish between these two cases is often difficult in practice. Recent work uses a combination of statistical models and algorithmic techniques to detect gene families that are under the effect of natural selection.

List of Gene Families

This is a list of gene families or gene complexes, that is sets of genes which occur across a number of different species which often serve similar biological functions. These gene families typically encode functionally related proteins, and sometimes the term gene families is a shorthand for the sets of proteins that the genes encode. They may or may not be physically adjacent on the same chromosome.

Genes-encoding Regulatory Proteins

- Homeobox (Hox gene family)
- Achaete-scute complex (Neuroblast formation)
- Krüppel-type zinc finger (ZNF)

Genes-encoding Immune System Proteins

- Major histocompatibility complex (MHC)

Related Protein Families

Motor Proteins

- Myosin
- Kinesin
- Dynein

Signal Transducing Proteins

- G-proteins
- MAP Kinase
- Olfactory Receptor
- Receptor tyrosine kinases

Transporters

- Cassidy proteins
- To be sorted and annotated

Source

- 14-3-3 family
- ABC Superfamily
- ABC Transporters
- ABC Transporters- PDR subfamily
- AGC Kinase Gene Family
- Aldehyde Dehydrogenase Gene Family
- Amino Acid/Auxin Permease AAAP Family
- Acyl Lipid Metabolism Gene Families
- Antiporters
- Aquaporins
- ARIADNE family
- AS2 family
- basic Helix-Loop-Helix Transcription Factor Gene Family
- Basic region leucine zipper (bZIP) Transcription Factor Gene Family
- Calcium Dependent Protein Kinase
- Calmodulin-binding Proteins
- Carbohydrate Esterase

- CBL-interacting serine-threonine protein kinases
- Cell Wall Biosynthesis
- Chloroplast and Mitochondria Gene Families
- COBRA Gene Family
- Core Cell Cycle Gene Families
- Cytochrome P450
- Cytochrome b5
- Cytoplasmic Ribosomal Protein Gene Family
- Cytoskeleton Gene Families
- DOF Gene Family
- EF-hand Containing Proteins
- Equilibrative Nucleoside Transporter Gene Family (ENT)
- Eukaryotic Initiation Factors Gene Family
- Expansin Gene Family
- F-Box Gene Family
- FH2 protein (formin) Gene Family
- FtsH Gene Family
- Glycoside Hydrolase
- Glycosyltransferase
- GRAS Gene Family
- Arabidopsis Glutathione S-transferase Gene Family
- Heat shock transcription factors Hsfs
- Histidine Kinase Gene Family
- Histidine Phosphotransfer Proteins
- IDZ Gene Family
- Inorganic Solute Cotransporters
- Ion Channels
- Lateral Organ Boundaries Gene Family
- Leucine-rich Repeat extensin Gene Families
- Lipid Metabolism Gene Families
- Kincsins gene family
- Major Intrinsic Protein MIP
- MAP Kinase Gene Family
- MAP Kinase Kinase Gene Family

- MAP Kinase Kinase Kinase
- MAP Kinase Kinase Kinase Kinase
- Miscellaneous Gene Families
- MLO family
- Monolignol Biosynthesis Gene Families
- Monosaccharide Transporter(-like) Gene Family
- MRS2 Gene Family
- MYB family
- Myosins gene family
- NADPH P450 reductases
- Nicotianamine Synthase Gene Family
- Nodulin-like Gene Family
- Nucleobase Ascorbate Transporter Gene Family
- Organic Solute Cotransporters
- Peroxidase, class III
- Phospholipase D Gene Family (Zarsky group)
- Phospholipase D Gene Family (Wang group)
- Plant Defensins Gene Families
- Polysaccharide Lyase
- Pollen Coat Proteome
- PP2C-type Phosphatase Gene Family
- Primary Pumps ATPases
- PRT Gene Family
- RAD5/RAD16-like Gene Family
- RCI2 Gene Family
- Rapid Response to Elicitors
- Receptor kinase-like Gene Family
- Response Regulator Gene Family
- SBP-box Gene Family
- SNARE and Associated Proteins
- Single Myb Histone (SMH) Gene Family
- Subtilisin-like Serine Gene Family
- Sucrose Transporter Gene Family
- Sulfurtransferasese/Rhodanese

- Strubbelig Receptor Gene Family
- rehalose Biosynthesis Gene Families
- ype I MADS-box Gene Family
- MADS like Gene Family
- MADS-box Gene Family
- U-box Gene Family
- Wox Gene Family
- WRKY Transcription Factor Superfamily
- Xyloglucan Fucosyltransferase1
- Zinc finger-homeobox gene family

Gene Patent

A gene patent is a patent on a specific isolated gene sequence, its chemical composition, the processes for obtaining or using it, or a combination of such claims. Gene patents are a part of the broader category of biological patents. Gene patents may claim the isolated natural sequences of genes, the use of a natural sequence for purposes such as diagnostic testing, or a natural sequence that has been altered by adding a promoter or other changes to make it more useful. The courts have upheld claims on altered sequences, but courts and lawyers' opinions have been mixed on upholding the use of natural sequences and particularly the sequence itself. Patents on genes have only been granted on isolated gene sequences with known functions, and these patents cannot be applied to the naturally occurring genes in humans or any other naturally occurring organism.

History

The United States has been patenting chemical compositions based upon human products for over 100 years. The first patent for a human product was granted on March 20, 1906 for a purified form of adrenaline. It was challenged and upheld in Parke-Davis v. Mulford Judge Hand argued that natural substances when they are purified are more useful than the original natural substances.

In 1980, the U.S. Supreme Court, in Diamond v. Chakrabarty, upheld the first patent on a newly-created living organism, a bacterium for digesting crude oil in oil spills. The patent examiner for the United States Patent and Trademark Office had rejected the patent of a living organism, but Chakrabarty appealed. As a rule, raw natural material is generally rejected for patent approval by the USPTO. The Court ruled that as long as the organism is truly "man-made," such as through genetic engineering, then it is patentable. Because the DNA of Chakrabarty's organism was modified, it was patentable.

Controversy

Law professor Rebecca S. Eisenberg argues that gene patents produce an "anticommons" at odds with an ideal scientific commons.

Others claim that patents have not created this "anticommons" effect on research, based on surveys of scientists.

Professional societies of pathologists have criticized patents on disease genes and exclusive licenses to perform DNA diagnostic tests. In the 2009 Myriad case, doctors and pathologists complained that the patent on BRCA1 and BRCA2 genes prevented patients from receiving second opinions on their test results. Pathologists complained that the patent prevented them from carrying out their medical practice of doing diagnostic tests on patient samples and interpreting the results.

Another example is a series of lawsuits filed by the Alzheimer's Institute of America (AIA) starting in 2010, concerning a gene patent it controls. The patent covers a genetic mutation that predisposes to Alzheimer's, and more importantly, it applies to transgenic mice carrying the mutation. These mice are widely used in Alzheimer's research, both by academic scientists doing basic research and by companies that use the mice to test products in development. Two of these suits are directed to companies that were started based on inventions made at universities (Comentis and Avid), and in each of those cases, the university has been sued along with the company. While none of the suits target universities that are conducting basic research using the mice, one of the suits is against Jackson Labs, a nonprofit company that provides transgenic mice to academic and commercial researchers and is an important repository of such mice.

While there is some controversy concerning the patenting of isolated genes and the way those patents are used, and there is controversy concerning patents on the diagnostic uses of genes (the real source of dispute in the Myriad court case), it is difficult to find controversy surrounding patents on genes that are used to manufacture therapeutic proteins (for an example of patents on therapeutic proteins, the drug candidate that is the subject of the early part of the movie 'Extraordinary Measures' was covered in part by a classic gene patent, US Patent 6,770,468.) There is also little controversy concerning the role of gene patents in the chemical industry—for instance in the manufacture of enzymes used in consumer products or industrial processes. In descriptions of the controversy over "gene patents", it is rare to find references to these everyday uses of genes and gene patents.

4

Myriad Genetics Case

In 2009, a lawsuit, Association For Molecular Pathology, et al., vs. United States Patent and Trademark Office, et al. was filed against Myriad Genetics and the U.S. Patent and Trademark Office by professional medical organizations, doctors, and patients, represented by the American Civil Liberties Union and the Public Patent Foundation of the Benjamin N. Cardozo School of Law, New York, in the U.S. District Court for the Southern District of New York

The complaint challenged specific claims on isolated genes and diagnostic methods in seven of Myriad's 23 patents on BRCA1 and BRCA2. Some of these patents are owned by Myriad and some of them were licensed from University of Utah. More specifically, the plaintiffs wanted certain claims declared invalid on the grounds that they are not patentable subject matter- that the isolated genes are unpatentable products of nature and that the diagnostic method claims are mere thought processes that do not yield any real world transformations.

Many patent lawyers had predicted that the courts would throw out this case. However, United States District Court Judge Robert W. Sweet accepted the case, and on March 29, 2010 ruled that the claims were invalid. Other claims, which were not challenged in the lawsuit, still stand.

Judge Sweet's 152–page decision ruled that the challenged claims to the isolated gene sequences had been "improperly granted" because they claimed unpatentable subject matter; the claims to the diagnostic methods were found invalid under the recent In re Bilski decision. Because the case could be decided with patent law, Judge Sweet did not look at the challenge on First

Amendment grounds and dismissed them without prejudice. However, on March 30, 2010, Myriad announced that it will appeal the decision. On June 16, 2010, Myriad filed its Notice of Appeal. Oral arguments were heard at the Federal Circuit on April 4, 2011.

On July 29, 2011 the Federal Circuit overturned the District court in part, and upheld the decision in part. The Federal Circuit overturned the District Court's finding that the claims covering isolated gene sequences are invalid and also overturned the invalidity of some of the diagnostic claims; the Federal Circuit upheld the finding that the claims for the diagnostic methods that only compare or analyze sequences - that have no transformative step - are invalid. Further appeals are likely; on October 12, 2011 the ACLU and the Public Patent Foundation said they intended to petition the Supreme Court to overturn the decision.

Gene Pool

In population genetics, a gene pool is the complete set of unique alleles in a species or population.

Description

A large gene pool indicates extensive genetic diversity, which is associated with robust populations that can survive bouts of intense selection. Meanwhile, low genetic diversity (see inbreeding and population bottlenecks) can cause reduced biological fitness and an increased chance of extinction, although as explained by genetic drift new genetic variants, that may cause an increase in the fitness of organisms, are more likely to fix in the population if it is rather small. When all individuals in a population are identical with regard to a particular phenotypic trait, the population is said to be monomorphic. When the individuals show several variants of a particular trait they are said to be polymorphic.

Gene Pool Concept in Crop Breeding

Harlan and de Wet (1971) proposed classifying each crop and its related species by gene pools rather than by formal taxonomy.

1. *Primary gene pool (GP-1):* Members of this gene pool are probably in the same "species" (in conventional biological usage) and can intermate freely. Harlan and de Wet wrote, "Among forms of this gene pool, crossing is easy; hybrids are generally fertile with good chromosome pairing; gene segregation is approximately normal and gene transfer is generally easy." They also advised subdividing each crop gene pool in two:
   - o Subspecies A: Cultivated races
   - o Subspecies B: Spontaneous races (wild or weedy)

2. *Secondary gene pool (GP-2):* Members of this pool are probably normally classified as different species than the crop species under consideration (the primary gene pool). However, these species are closely related and can cross and produce at least some fertile hybrids (F1s). As would be expected by members of different species, there are some reproductive barriers between members of the primary and secondary gene pools:
   - o hybrids may be weak
   - o hybrids may be partially sterile
   - o chromosomes may pair poorly or not at all
   - o recovery of desired phenotypes may be difficult in subsequent generations
   - o However, "The gene pool is available to be utilized, however, if the plant breeder or geneticist is willing to put out the effort required."
3. *Tertiary gene pool (GP-3):* Members of this gene pool are more distantly related to the members of the primary gene pool. The primary and tertiary gene pools can be intermated, but gene transfer between them is impossible without the use of "rather extreme or radical measures" such as:
   - o embryo rescue (or embryo culture, a form of plant organ culture)
   - o induced polyploidy (chromosome doubling)
   - o bridging crosses (e.g., with members of the secondary gene pool).

Gene Flow

In population genetics, gene flow (also known as gene migration) is the transfer of alleles of genes from one population to another. Migration into or out of a population may be responsible for a marked change in allele frequencies (the proportion of members carrying a particular variant of a gene). Immigration may also result in the addition of new genetic variants to the established gene pool of a particular species or population.

There are a number of factors that affect the rate of gene flow between different populations. One of the most significant factors is mobility, as greater mobility of an individual tends to give it greater migratory potential. Animals tend to be more mobile than plants, although pollen and seeds may be carried great distances by animals or wind. Maintained gene flow between two populations can also lead to a combination of the two gene pools, reducing the genetic variation between the two groups. It is for this reason that gene flow strongly acts against speciation, by recombining the gene pools of the groups, and thus, repairing the developing differences in genetic variation that would have led to full speciation and creation of daughter species.

For example, if a species of grass grows on both sides of a highway, pollen is likely to be transported from one side to the other and vice versa. If this pollen is able to fertilize the plant where it ends up and produce viable offspring, then the alleles in the pollen have effectively been able to move from the population on one side of the highway to the other.

Barriers to Gene Flow

Physical barriers to gene flow are usually, but not always, natural. They may include impassable mountain ranges, oceans, or vast deserts. In some cases, they can be artificial, man-made barriers, such as the Great Wall of China, which has hindered the gene flow of native plant populations. One of these native plants, Ulmus pumila, demonstrated a lower prevalence of genetic differentiation than the plants Vitex negundo, Ziziphus jujuba, Heteropappus hispidus, and Prunus armeniaca whose habitat is located on the opposite side of the Great Wall of China where Ulmus pumila grows.

This is because Ulmus pumila has wind-pollination as its primary means of propagation and the latter-plants carry out pollination through insects. Samples of the same species which grow on either side have been shown to have developed genetic differences, because there is little to no gene flow to provide recombination of the gene pools.

Barriers to gene flow need not always be physical. Species can live in the same environment, yet show very limited gene flow due to limited hybridization or hybridization yielding unfit hybrids.

Gene Flow in Humans

Gene flow has been observed in humans. For example, in the United States, gene flow was observed between a white European population and a black West African population, which were recently brought together. In West Africa, where malaria is prevalent, the Duffy antigen provides some resistance to the disease, and this allele is thus present in nearly all of the West African population. In contrast, Europeans have either the allele Fy or Fy because malaria is almost non-existent. By measuring the frequencies of the West African and European groups, scientists found that the allele frequencies became mixed in each population because of movement of individuals. It was also found that this gene flow between European and West African groups is much greater in the Northern U.S. than in the South.

Gene Flow Between Species

Gene flow can occur between species, either through hybridization or gene transfer from bacteria or virus to new hosts. Gene transfer, defined as the movement of genetic material across species boundaries, which includes horizontal gene transfer, antigenic shift, and reassortment is sometimes an important source of genetic variation. Viruses can transfer genes between species. Bacteria can incorporate genes from other dead bacteria, exchange genes with living bacteria, and can exchange plasmids across species boundaries. "Sequence comparisons suggest recent horizontal transfer of many genes among diverse species including across the boundaries of phylogenetic "domains". Thus determining the phylogenetic history of a species can not be done conclusively by determining evolutionary trees for single genes."

Biologist Gogarten suggests "the original metaphor of a tree no longer fits the data from recent genome research". Biologists [should] instead use the metaphor of a mosaic to describe the different histories combined in individual genomes and use the metaphor of an intertwined net to visualize the rich exchange and cooperative effects of horizontal gene transfer.

"Using single genes as phylogenetic markers, it is difficult to trace organismal phylogeny in the presence of HGT [horizontal gene transfer]. Combining the simple coalescence model of cladogenesis with rare HGT [horizontal gene transfer] events suggest there was no single last common ancestor that contained all of the genes ancestral to those shared among the three domains of life. Each contemporary molecule has its own history and traces back to an individual molecule cenancestor. However, these molecular ancestors were likely to be present in different organisms at different times."

Genetic Pollution

Purebred, naturally-evolved, region-specific, wild species can be threatened with extinction through the process of genetic pollution, potentially causing uncontrolled hybridization, introgression and genetic swamping. These processes can lead to homogenization or replacement of local genotypes as a result of either a numerical and/or fitness advantage of introduced plant or animal. Nonnative species can bring about a form of extinction of native plants and animals by hybridization and introgression either through purposeful introduction by humans or through habitat modification, bringing previously isolated species into contact.

These phenomena can be especially detrimental for rare species coming into contact with more abundant ones. Interbreeding between the species can cause a 'swamping' of the rarer species' gene pool, creating hybrids that drive the originally purebred native stock to complete extinction.

The extent of this facet of gene flow is not always apparent from morphological (outward appearance) observations alone. Some degree of gene flow may be due to normal, evolutionarily constructive processes, and all constellations of genes and genotypes cannot be preserved. That being said, hybridization with or without introgression may threaten a rare species' existence nonetheless.

Models of Gene Flow

Models of gene flow can be derived from population genetics, e.g. Sewall Wright's neighborhood model, Wright's island model and the stepping stone model.

Gene Flow Mitigation

This section does not cite any references or sources. Please help improve this section by adding citations to reliable sources. Unsourced material may be challenged and removed. (March 2010)

When cultivating genetically modified (GM) plants or livestock, it becomes necessary to prevent "genetic pollution" i.e. their genetic modification from reaching other conventionally hybridized or wild native plant and animal populations by using gene flow mitigation usually through unintentional cross pollination and crossbreeding. Reasons to limit gene flow may include biosafety or agricultural co-existence, in which GM and non-GM cropping systems work side by side. Scientists in several large research programmes are investigating methods of limiting gene flow in plants. Among these programmes are Transcontainer, which investigates methods for biocontainment, SIGMEA, which focuses on the biosafety of genetically modified plants, and Co-Extra, which studies the co-existence of GM and non-GM product chains.

Generally, there are three approaches to gene flow mitigation: keeping the genetic modification out of the pollen, preventing the formation of pollen, and keeping the pollen inside the flower.

- The first approach requires transplastomic plants. In transplastomic plants, the modified DNA is not situated in the cell's nucleus but is present in plastids, which are cellular compartments outside the nucleus. An example for plastids are chloroplasts, in which photosynthesis occurs. In some plants, the pollen does not contain plastids and, consequently, any modification located in plastids cannot be transmitted by the pollen.
- The second approach relies on male sterile plants. Male sterile plants are unable to produce functioning flowers and therefore cannot release viable pollen. Cytoplasmic male sterile plants are known to produce higher yields. Therefore, researchers are trying to introduce this trait to genetically modified crops.
- The third approach works by preventing the flowers from opening. This trait is called cleistogamy and occurs naturally in some plants. Cleistogamous plants produce flowers which either open only partly or not at all. However, it remains unclear how reliable cleistogamy is for gene flow mitigation: a Co-Extra research project on rapeseed investigating the matter has published preliminary results which cast doubt on the attainment of a high degree of reliability.

Biological Dispersal

Biological dispersal refers to species movement away from an existing population or away from the parent organism. Through simply moving from one habitat patch to another, the dispersal of an individual has consequences not only for individual fitness, but also for population dynamics, population genetics, and species distribution.

Understanding dispersal and the consequences both for evolutionary strategies at a species level, and for processes at an ecosystem level, requires understanding on the type of dispersal, the dispersal range of a given species, and the dispersal mechanisms involved.

Biological dispersal may be contrasted with geodispersal, which is the mixing of previously isolated populations (or whole biotas) following the erosion of geographic barriers to dispersal or gene flow (Lieberman, 2005;Albert and Reis, 2011).

Types of Dispersal

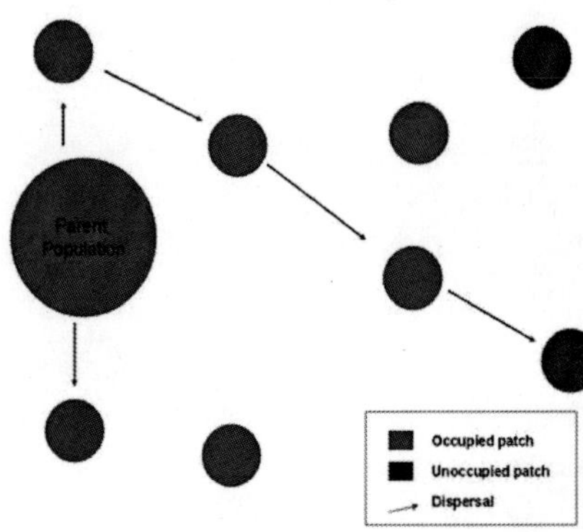

Figure: *Dispersal from parent population*

At some time during its life, an organism, whether animal or plant, moves, or is moved, so that it or its offspring do not die exactly where they were born. Such movement is called dispersal. Some organisms are motile throughout their lives, but others are adapted to move or be moved at precise, limited phases of their life cycles. This is commonly called the dispersive phase of the life cycle. The strategies of organisms' entire life cycles often are predicated on the nature and circumstances of their dispersive phases.

Density Independent Dispersal

Organisms have evolved adaptations for dispersal that take advantage of various forms of kinetic energy occurring naturally in the environment. This is referred to as density independent or passive dispersal and operates on many groups of organisms (some invertebrates, fish, insects and sessile organisms such as plants) that depend on animal vectors, wind, gravity or current for dispersal.

Density Dependent Dispersal

Density dependent or active dispersal for many animals largely depends on factors such as local population size, resource competition, habitat quality, and habitat size.

Due to population density, dispersal may relieve pressure for resources in an ecosystem, and competition for these resources may be a selection factor for dispersal mechanisms. Dispersal of organisms is a critical process for understanding both geographic isolation in evolution through gene flow and the broad patterns of current geographic distributions (biogeography).

Dispersal Range

"Dispersal range" refers to the distance a species can move from an existing population or the parent organism. An ecosystem depends critically on the ability of individuals and populations to disperse from one habitat patch to another. Therefore, biological dispersal is critical to the stability of ecosystems.

Environmental Constraints

Few species are ever evenly or randomly distributed within or across landscapes. In general, species significantly vary across the landscape in association with environmental features that influence their reproductive success and population persistence. Spatial patterns in environmental features (e.g. resources) permit individuals to escape unfavorable conditions and seek out new locations.

This allows the organism to "test" new environments for their suitability, provided they are within animal's geographic range. In addition, the ability of a species to disperse over a gradually changing environment could enable a population to survive extreme conditions. (i.e. climate change).

Dispersal Barriers

A dispersal barrier may mean that the dispersal range of a species is much smaller than the species distribution. An artificial example is habitat fragmentation due to human land use. Natural barriers to dispersal that limit species distribution include mountain ranges and rivers.

An example is the separation of the ranges of the two species of chimpanzee by the Congo River.

On the other hand, human activities may also expand the dispersal range of a species by providing new dispersal methods (e.g., ships).

Dispersal Mechanisms

Most animals are capable of locomotion and the basic mechanism of dispersal is movement from one place to another. Locomotion allows the organism to "test" new environments for their suitability, provided they are within the animal's range. Movements are usually guided by inherited behaviors.

The formation of barriers to dispersal or gene flow between adjacent areas can isolate populations on either side of the emerging divide. The geographic separation and subsequent genetic isolation of portions of an ancestral population can result in speciation.

Plant Dispersal Mechanisms

Figure: *Burs are an example of a seed dispersion mechanism which uses a biotic vector, in this case animals with fur.*

Seed dispersal is the movement or transport of seeds away from the parent plant. Plants have limited mobility and consequently rely upon a variety of dispersal vectors to transport their propagules, including both abiotic and biotic vectors. Seeds can be dispersed away from the parent plant individually or collectively, as well as dispersed in both space and time. The patterns of seed dispersal are determined in large part by the dispersal mechanism and this has important implications for the demographic and genetic structure of plant populations, as well as migration patterns and species interactions. There are five main modes of seed dispersal: gravity, wind, ballistic, water and by animals.

Animal Dispersal Mechanisms

Non-motile Animals

There are numerous animal forms that are non—motile, such as sponges, bryozoans, tunicates, sea anemones, corals, and oysters. In common, they are all either marine or aquatic. It may seem curious that plants have **been so** successful at stationary life on land, while animals have not, but the answer lies in the food supply. Plants produce their own food from sunlight and carbon dioxide—both generally more abundant on land than in water. Animals fixed in place must rely on the surrounding medium to bring food at least close enough to grab, and this occurs in the three-dimensional water environment, but with much less abundance in the atmosphere.

All of the marine and aquatic invertebrates whose lives are spent fixed to the bottom (more or less; anemones are capable of getting up and moving to a new location if conditions warrant) produce dispersal units. These may be specialized "buds", or motile sexual reproduction products, or even a sort of alteration of generations as in certain cnidaria.

Corals provide a good example of how sedentary species achieve dispersion. Corals reproduce by releasing sperm and eggs directly into the water. These release events are coordinated by lunar phase in certain warm months, such that all corals of one or many species on a given reef will release on the same single or several consecutive nights. The released eggs are fertilized, and the resulting zygote develops quickly into a multicellular planula. This motile stage then attempts to find a suitable substratum for settlement. Most are unsuccessful and die or are fed upon by zooplankton and bottom dwelling predators such as anemones and other corals. However, untold millions are produced, and a few do succeed in locating spots of bare limestone, where they settle and transform by growth into a polyp. All things being favorable, the single polyp grows into a coral head by budding off new polyps to form a colony.

Motile Animals

The majority of all animals are motile. Although motile animals can, in theory, disperse themselves by their spontaneous and independent locomotive powers, a great many species utilize the existing kinetic energies in the environment, resulting in passive movement. Dispersal by water currents is especially associated with the physically small inhabitants of marine waters known as zooplankton. The term plankton comes from the Greek, ðëáãêôïí, meaning "wanderer" or "drifter".

Dispersal by Dormant Stages

Many animal species, especially freshwater invertebrates, are able to disperse by wind or by transfer with an aid of larger animals (birds, mammals or fishes) as dormant eggs, dormant embryos or, in some cases, dormant adult stages. Tardigrades, some rotifers and some copepods are able to withstand desiccation as adult dormant stages. Many other taxa (Cladocera, Bryozoa, Hydra, Copepoda and so on) can disperse as dormant eggs or embryos. Freshwater sponges usually have special dormant propagules called gemmulae for such a dispersal. Many kinds of dispersal dormant stages are able to withstand not only desiccation and low and high temperature, but also action of digestive enzymes during their transfer through digestive tracts of birds and other animals, high concentration of salts and many kinds of toxicants. Such dormant-resistant stages made possible the long-distance dispersal from one water body to another and broad distribution ranges of many freshwater animals.

Genetic Erosion

Genetic erosion is a process whereby an already limited gene pool of an endangered species of plant or animal diminishes even more when individuals from the surviving population die off without getting a chance to meet and breed with others in their endangered low population.

Genetic erosion occurs because each individual organism has many unique genes which get lost when it dies without getting a chance to breed. Low genetic diversity in a population of wild animals and plants leads to a further diminishing gene pool, inbreeding and a weakening immune system and fast tracks that species towards eventual extinction. All the world's endangered species are plagued by varying degrees of genetic erosion and most need a human assisted breeding program to keep their population viable and to keep them from going extinct in the long run.

The smaller the population is, the more magnified the effect of genetic erosion gets when each surviving individual of the species is lost without getting a chance to breed. Genetic erosion gets compounded and accelerated by habitat fragmentation, today most endangered species live in smaller and smaller chunks of fragmented habitat interspersed with human settlements and farmland making it impossible for them to naturally meet and breed with others of their kind, many die off without getting a chance to breed and pass on their genes in the living population. The gene pool of a species or a population is the complete set of unique alleles that would be found by inspecting the genetic material of every living member of that species or population. A large gene pool indicates extensive genetic diversity, which is associated with robust populations that can survive bouts of intense selection. Meanwhile, low genetic diversity (see inbreeding and population bottlenecks) can cause reduced biological fitness and an increased chance of extinction.

Processes and Consequences

"A population bottleneck creates a shrinking gene pool that leaves fewer and fewer mating partners. What are the genetic implications? The animals become part of a high stakes poker game — with a crooked dealer. After beginning with a 52-card deck, the players wind up with, say, five cards that they are dealt over and over. As they begin to inbreed, congenital effects appear, both physical and reproductive. Often abnormal sperm increase; infertility rises; the birthrate falls. Most perilous in the long run, each animal's immune defense system is weakened. Thus, even if an endangered species in a bottleneck can withstand whatever human development may be eating away at its habitat, it still faces the threat of an epidemic that could well be fatal to the entire population."

Genetic Erosion in Agricultural and Livestock Biodiversity

Genetic erosion in agricultural and livestock biodiversity is the loss of genetic diversity, including the loss of individual genes, and the loss of particular combinants of genes (or gene complexes) such as those manifested in locally adapted landraces of domesticated animals or plants adapted to the natural environment in which they originated. The term genetic erosion is sometimes used in a narrow sense, such as for the loss of alleles or genes, as well as more broadly, referring to the loss of varieties or even species. The major driving forces behind genetic erosion in crops are: variety replacement, land clearing, overexploitation of species, population pressure, environmental degradation, overgrazing, policy and changing agricultural systems. The main factor, however, is the replacement of local varieties of domestic plants and animals by other varieties or species. A large number of varieties can also often be dramatically reduced when commercial varieties are introduced into traditional farming systems. Many researchers believe that the main problem related to agro-ecosystem management is the general tendency towards genetic and ecological uniformity imposed by the development of modern agriculture. Human intervention, modern science and safeguards to guard against genetic erosion

In-situ Conservation

With advances in modern science several techniques and safeguards have emerged to check the relentless advance of genetic erosion and the resulting acceleration of endangered species towards extinction. However many of these techniques and safeguards are too expensive yet to be practical, the best way to protect species is to protect their habitat and to let them live in it naturally. Wildlife sanctuaries and national parks have been created to preserve entire ecosystems with all the web of species which call them home. Wildlife corridors are created to join fragmented habitats (Habitat fragmentation) to enable endangered species to travel, meet and breed with others of their kind. Scientific conservation and modern wildlife management techniques with the help of scientifically trained staff help manage these protected ecosystems and the wildlife found in them. Wild animals are also translocated and reintroduced to other locations physically when fragmented wildlife habitat is too far and isolated to be able to link it with a wildlife corridor or when local extinction has already occurred.

Ex-situ Conservation

Modern policies of the zoo associations and zoos around the world have changed to putting extreme importance on keeping and breeding wild sourced species and subspecies of animals in their registered endangered species breeding programs which will have a chance to be reintroduced and survive in the wild. The main objectives of zoos today have changed to breed species and subspecies to assist conservation efforts in the wild. Zoos do this by maintaining extremely detailed scientific breeding records (i.e., studbooks) and loaning their wild animals to other zoos around the country and indeed globally for breeding to safeguard against inbreeding.

Costly and sometimes controversial ultra modern ex-situ conservation techniques have emerged for saving the genetic biodiversity on our planet and the diversity in their gene pool by guarding against genetic erosion through modern concepts like seedbanks, sperm banks and tissue banks. Sperms, eggs and embryos can now be frozen and kept in these banks which are sometimes called Modern Noah's Ark or Frozen Zoos. Cryopreservation techniques are used to freeze these living materials and keep them alive by storing them submerged in liquid nitrogen tanks.

Thus, preserved material can then be used for artificial insemination, in vitro fertilization, embryo transfer and cloning to protect diversity in the gene pool of critically endangered species. It is today possible to save endangered species from extinction by preserving parts like tissue, sperms, eggs etc. even after the death of a critically endangered animal or collected from one found freshly dead in captivity or from the wild and resurrect it with the help of cloning and give it another chance to breed its genes into the living population of the respective species which is threatened with extinction. Resurrection of dead critically endangered wildlife with the help of cloning is still being perfected and is still too expensive to be practical but with time and advancement is science it may well become a routine procedure. Recently strategies have been made to find an integrated approach to in situ and ex situ conservation.

Genetic Admixture

Genetic admixture occurs when individuals from two or more previously separated populations begin interbreeding. Admixture results in the introduction of new genetic lineages into a population. It has been known to slow local adaptation by introducing foreign, unadapted genotypes (known as gene swamping).

It also prevents speciation by homogenizeing populations. Genetic admixture often occurs when a geographic barrier separating populations,such as a river or isthmus, is removed or when anthropogenic activities result in movement of populations (for example invasive species). One example of genetic admixture resulting from the introduction of an invasive species is provided by the Cuban brown anole. Several isolated populations of this species exist in the native range of Cuba. However in the introduced range of Florida, these populations freely interbreed, forming an admixed population.

Miscegenation

Miscegenation (Latin miscere "to mix" + genus "kind") is the mixing of different racial groups through marriage, cohabitation, sexual relations, and procreation.

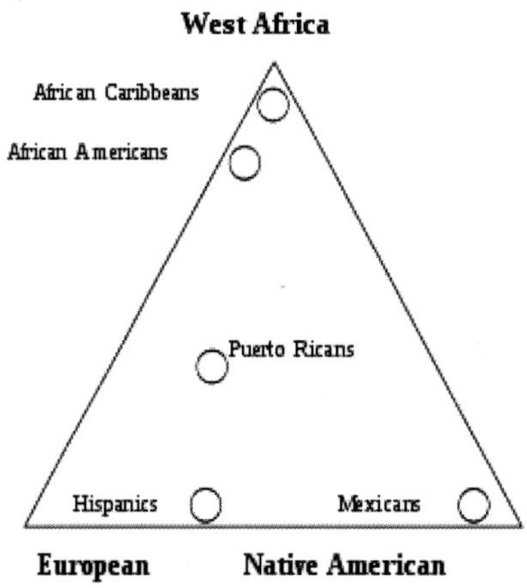

Figure: *Triangle plot shows average admixture of five North American ethnic groups. Individuals that self-identify with each group can be found at many locations on the map, but on groups tend to cluster differently.*

The term miscegenation has been used since the 19th century to refer to interracial marriage and interracial sex, and more generally to the process of racial admixture, which has taken place since ancient history. The term entered historical records during European colonialism and the Age of Discovery, but societies such as China and Japan also had restrictions on marrying with peoples they considered different. Historically the term has been used in the context of laws banning interracial marriage and sex, so-called anti-miscegenation laws. It is a potentially offensive word.

Usage

Today, the word miscegenation is avoided by many scholars, because the term suggests an actual biological phenomenon, rather than its nature as a categorization imposed on certain relationships. The word is considered offensive by many, and other terms such as "interracial", "interethnic" or "cross-cultural" are more common in contemporary usage. The term remains in use among scholars when referring to past practices concerning multiraciality, such as anti-miscegenation laws that banned interracial marriages. In Spanish, Portuguese and French, the words used to describe the mixing of races are mestizaje, mestiçagem and métissage. These words, much older than the term miscegenation, are derived from the Late Latin mixticius for "mixed", which is also the root of the Spanish word mestizo. Portuguese also uses miscigenação, derived from the same Latin root as the English word.

These non-English terms for "race-mixing" are not considered as offensive as "miscegenation", although they have historically been tied to the caste system (Casta) that was established during the colonial era in Spanish-speaking Latin America. Some groups in South America, however, consider the use of the word mestizo offensive because it was used during the times of the colony to refer specifically to the mixing between the conquistadores and the indigenous people.

Today, the mixes among races and ethnicities are diverse, so it is considered preferable to use the term "mixed-race" or simply "mixed" (mezcla). In Portuguese-speaking Latin America (i.e., Brazil), a milder form of caste system existed, although it also provided for legal and social discrimination among individuals belonging to different races, since slavery for blacks existed until the late 19th century.

Miscegenation occurred significantly from the very first settlements, with their descendants achieving high rank in government and society. To this day, the Brazilian class system is drawn mostly around socio-economic lines, not racial ones (in a manner similar to other former Portuguese colonies).

The concept of miscegenation is tied to concepts of racial difference. As the different connotations and etymologies of miscegenation and mestizaje suggest, definitions of race, "race mixing" and multiraciality have diverged globally as well as historically, depending on changing social circumstances and cultural perceptions.

Mestizo are people of mixed white and indigenous, usually Amerindian ancestry, who do not self-identify as indigenous peoples or Native Americans. In Canada, however, the Métis, who also have partly Amerindian and partly white, often French-Canadian, ancestry, have identified as an ethnic group and are a constitutionally recognized aboriginal people. (In Canada these are referred to as métis. A very famous one was Louis Riel.)

The differences between related terms and words which encompass aspects of racial admixture show the impact of different historical and cultural factors leading to changing social interpretations of race and ethnicity. Thus the Comte de Montlosier, in exile during the French Revolution, equated class difference in 18th-century France with racial difference. Borrowing Boulainvilliers' discourse on the "Nordic race" as being the French aristocracy that invaded the plebeian "Gauls", he showed his contempt for the lowest social class, the Third Estate, calling it "this new people born of slaves... mixture of all races and of all times".

Etymological History

Miscegenation comes from the Latin miscere, "to mix" and genus, "kind". The word was coined in the U.S. in 1863, and the etymology of the word is tied up with political conflicts during the American Civil War over the abolition of slavery and over the racial segregation of African-Americans. The reference to genus was made to emphasize the supposedly distinct biological differences between whites and non-whites, though all humans belong to the same genus, Homo, same species, Homo sapiens and same subspecies, Homo sapiens sapiens.

The word was coined in an anonymous propaganda pamphlet published in New York City in December 1863, during the American Civil War. The pamphlet was entitled Miscegenation: The Theory of the Blending of the Races, Applied to the American White Man and Negro. It purported to advocate the intermarriage of whites and blacks until they were indistinguishably mixed, as a desirable goal, and further asserted that this was the goal of the Republican Party. The pamphlet was a hoax, concocted by Democrats, to discredit the Republicans by imputing to them what were then radical views that offended against the attitudes of the vast majority of whites, including those who opposed slavery. There was already much opposition to the war effort; in New York in particular the opposition reached heights of the Draft Riots, that included numerous lynchings.

The pamphlet and variations on it were reprinted widely in both the north and south by Democrats and Confederates. Only in November 1864 was the pamphlet exposed as a hoax. The hoax pamphlet was written by David Goodman Croly, managing editor of the New York World, a Democratic Party paper, and George Wakeman, a World reporter.

By then, the word miscegenation had entered the common language of the day as a popular buzzword in political and social discourse. The issue of miscegenation, raised by the opponents of Abraham Lincoln, featured prominently in the election campaign of 1864.

In the United States, miscegenation has referred primarily to the intermarriage between whites and non-whites, especially blacks. Before the publication of Miscegenation, the word amalgamation, borrowed from metallurgy, had been in use as a general term for ethnic and racial intermixing. A contemporary usage of this metaphor was Ralph Waldo Emerson's private vision in 1845 of America as an ethnic and racial smelting-pot, a variation on the concept of the melting pot. Opinions in the U.S on the desirability of such intermixing, including that between white Protestants and Irish Catholic immigrants, were divided. The term miscegenation was coined to refer specifically to the intermarriage of blacks and whites, with the intent of galvanising opposition to the war.

The Concept of Miscegenation

Franz Boas (considered to be the 'father' of American cultural anthropology) as well as many of his students, such as Ashley Montagu, considered race to be an invalid concept. From this point of view, if the concept of race is invalid, then miscegenation as the 'crossing' of races is equally invalid. These ideas are pursued in greater depth in Ashley Montagu's books on the subject.

Laws Banning Miscegenation

Laws banning "race-mixing" were enforced in Nazi Germany (the Nuremberg Laws) from 1935 until 1945, in certain U.S. states from the Colonial era until 1967 and in South Africa during the early part of the Apartheid era. All these laws primarily banned marriage between spouses of different racially or ethnically defined groups, which was termed "amalgamation" or "miscegenation" in the U.S. The laws in Nazi Germany and many of the U.S. states, as well as South Africa, also banned sexual relations between such individuals. In the United States, the various state laws prohibited the marriage of whites and blacks, and in many states also the intermarriage of whites with Native Americans or

Asians. In the U.S., such laws were known as anti-miscegenation laws. From 1913 until 1948, 30 out of the then 48 states enforced such laws. Although an "Anti-Miscegenation Amendment" to the United States Constitution was proposed in 1871, in 1912–1913, and in 1928, no nation-wide law against racially mixed marriages was ever enacted. In 1967, the United States Supreme Court unanimously ruled in Loving v. Virginia that anti-miscegenation laws are unconstitutional. With this ruling, these laws were no longer in effect in the remaining 16 states that still had them.

The laws in U.S. states were established to maintain "racial purity" and white supremacy. Such laws were passed in South Africa because of fears that the white minority would be "bred-out" by the black majority.

The Nazi ban on interracial marriage and interracial sex was enacted in September 1935 as part of the Nuremberg Laws, the Gesetz zum Schutze des deutschen Blutes und der deutschen Ehre (The Law for the Protection of German Blood and German Honour). The Nuremberg Laws classified Jews as a race, and forbade marriage and extramarital sexual relations between persons of Jewish origin and persons of "German or related blood". Such intercourse was condemned as Rassenschande (lit. "race-disgrace") and could be punished by imprisonment (usually followed by deportation to a concentration camp) and even by death. The Prohibition of Mixed Marriages Act in South Africa, enacted in 1949, banned intermarriage between different racial groups, including between whites and non-whites. The Immorality Act, enacted in 1950, also made it a criminal offense for a white person to have any sexual relations with a person of a different race. Both laws were repealed in 1985.

History of ethnoracial admixture and attitudes towards miscegenation

Africa

Africa has a long history of interracial mixing with Arabs and later European men having sexual relations with black African women as well as taking them as wives. Arabs played a big role in the African slave trade and unlike the trans-atlantic trade most of the black African slaves in the Arab slave trade were women. Most of them were used as sexual slaves by the Arab men and some were even taken as wives.

Sir Richard Francis Burton writes during his expedition to Africa about relationships between black women and white men. He writes, "The women are well disposed toward strangers of fair complexion, apparently with the permission of their husbands." There are several mulatto populations throughout Africa mostly the results of interracial relationships between Arab and European men and black women. In South Africa there are big mulatto communities like the Coloureds and Griqua formed by White colonists taking native African wives. In Namibia there is a community called the Rehoboth Basters formed by the interracial marriage of Dutch/German men and black African women.

In the former Portuguese Africa (now known as Angola, Mozambique and Cape Verde) racial mixing between white Portuguese and black Africans was fairly common, especially in Cape Verde where the majority of the population is of mixed descent.

There have been some recorded cases of Chinese merchants and labourers taking African wives throughout Africa as many Chinese workers were employed to build railways and other infrastructural projects in Africa. These labour groups were made up completely of men with very few Chinese women coming to Africa.

In West Africa, especially Nigeria there are many cases of Lebanese men taking African women. Many such mixed people have gained prominent positions in Africa. Flight Lieutenant Jerry John Rawlings, who has a Scottish father and a black Ghanaian mother became the President and Head of State of Ghana. Jean Ping, the son of a Chinese trader and a black Gabonese mother, became the Deputy Prime Minister as well as the Foreign minister of Gabon and is currently the Chairperson of the Commission of the African Union in 2009. Ian Khama son of a Botswanian leader and a white mother is the

President of Botswana and the paramount chief of the Bamangwato people. Indian men, who have long been traders in East Africa, at times married among local African women. The British Empire brought many Indian workers into East Africa to build the Uganda Railway. Indians eventually populated South Africa, Kenya, Uganda, Tanzania, Malawi, Rwanda, Zambia, Zimbabwe and Zaire in small numbers. These interracial unions were mostly unilateral marriages between Indian men and East African women.

Réunion

The majority of the population of Réunion is defined as mixed race. In the last 350 years, various ethnic groups (Africans, Chinese, English, French, Gujuratis, Tamil Indians) have arrived and settled on the island. There have been mixed race people on the island since its first permanent inhabitation in 1665.

Americas

Canada

The 2006 Census counted 289,400 mixed unions (marriages and common-law unions) involving a visible minority person with a non-visible minority person or a person from a different visible minority group. This was a 33.1% increase from 2001. Among all mixed unions in 2006, 247,600 couples were in unions involving a visible minority person and someone who was not a visible minority. The groups most likely to be involved in a mixed union were Japanese (74.4%), Latin Americans (47%) and blacks (40.6%).

Jamaica and Haiti

By some estimates, 80,000 North American and European women (most of them over the age of 40) visit Jamaica and Haiti every year for sex with young men (mostly in their 20s). They're called "milk bottles". HIV/AIDS infection rates in the Caribbean are much higher than in Canada or the U.S. Even so, female sex tourists in the Caribbean are not especially preoccupied by the risk.

United States

Historically, "race mixing" between black and white people was taboo in the United States. So-called anti-miscegenation laws, barring blacks and whites from marrying or having sex, were established in colonial America as early as 1691. The 1691 Virginia law was amended in 1705 to remove Indian-white intermarriage from the prohibition. Thomas Jefferson's policy proposal for dealing with Native Americans was "to let our settlements and theirs meet and blend together, to intermix, and become one people." He regretted in 1813 that white-Indian war had prevented this: "They would have mixed their blood with ours, and been amalgamated and identified with us within no distant period of time." The taboo among American whites surrounding white-black intermarriage can be seen as a historical consequence of the oppression and racial segregation of African-Americans. In many U.S. states interracial marriage was already illegal when the term miscegenation was invented in 1863. The first laws banning interracial marriage were introduced in the late 17th century in the slave-holding colonies of Virginia (1691) and Maryland (1692). Later these laws also spread to colonies and states where slavery did not exist.

It has also been argued that the first laws banning interracial marriage were a response by the planter elite to the problems they were facing due to the socio-economic dynamics of the plantation system in the Southern colonies. The bans in Virginia and Maryland were established at a time when slavery was not yet fully institutionalized. At the time, most forced laborers on the plantations were indentured servants, and they were mostly white. Some historians have suggested that the at-the-time unprecedented laws banning interracial marriage were originally invented by planters as a divide and rule tactic after the uprising of servants in Bacon's Rebellion. According to this theory, the ban on interracial marriage was issued to split up the racially mixed, increasingly mixed-race labour force into whites, who were given their freedom, and blacks, who were later treated as slaves rather than as indentured servants. By forbidding interracial marriage, it became possible to keep these two new groups separated and prevent a new rebellion.

In 1918, there was considerable controversy in Arizona when an Indian farmer B. K. Singh married the sixteen year-old daughter of one of his white tenants. During and after slavery, most American whites regarded interracial marriage between whites and blacks as taboo. However, during slavery many white American men and women did conceive children with black partners. These children automatically became slaves if the mother was a slave or were born free if the mother was free, as slavery was matrilineal. Some children were freed by their slave-holding fathers or bought to be emancipated if the father was not the owner. Many children of these unions formed enclaves under names such as Colored and Gens de couleur, etc. Most mixed-raced descendants merged into the African-American ethnic group during Jim Crow, while over the centuries a minority of mixed-raced Americans passed and became white, and others exist to this day in small mixed enclaves of Mestees such as the Melungeons and Lumbee.

Genetic research suggests that a considerable minority of white Americans (estimated at 1/3 of the population by some geneticists such as Mark Shriver) has some distant African-American ancestry, and that the majority of black Americans have some European ancestry. After the Civil War and the abolition of slavery in 1865, the intermarriage of white and black Americans continued to be taboo, especially but not only in the former slave states. The Motion Picture Production Code of 1930, also known as Hays Code, explicitly stated that the depiction of "miscegenation... is forbidden." One important strategy intended to discourage the marriage of white Americans and Americans of partly African descent was the promulgation of the one-drop theory, which held that any person with any known African ancestry, however remote, must be regarded as "black". This definition of blackness was encoded in the anti-miscegenation laws of various U.S. states, such as Virginia's Racial Integrity Act of 1924. The plaintiffs in Loving v. Virginia, Mildred Jeter and Richard Loving became the historically most prominent interracial couple in the US through their legal struggle against this act.

Accusations of support for miscegenation were commonly made by slavery defenders against Abolitionists before the Civil War. After the War, similar charges were used by white segregationists against advocates of equal rights for African Americans. They were said to be secretly plotting the destruction of the white race through miscegenation. In the 1950s, segregationists alleged a Communist plot funded by the Soviet Union with that goal. In 1957, segregationists cite the anti-semitic hoax A Racial Program for the Twentieth Century as evidence for these claims.

In 1958, the Christian fundamentalist preacher Jerry Falwell, at the time a defender of segregation, in a sermon railed against integration, warning that it would lead to miscegenation, which would "destroy our [white] race eventually.".

Asians were also specifically included in some state laws. California continued to ban Asian/white marriages until the Perez v. Sharp decision in 1948.

In the United States, segregationists and Christian identity groups have claimed that several passages in the Bible, for example the stories of Phinehas and of the so-called "curse of Ham", should be understood as referring to miscegenation and that certain verses expressly forbid it. Most theologians read these verses and references as forbidding inter-religious marriage, rather than inter-racial marriage.

Miscegenation has become increasingly accepted in the United States since the Civil Rights movement and up to the present day. The most notable American of mixed race is the current President of the United States, Barack Obama, who is the product of a "mixed" marriage between a black father and white mother. Nevertheless, as late as 2009, a Louisiana justice of the peace refused to issue a marriage license to an interracial couple, justifying the decision on grounds of concern for any children the couple might have.

In recent years, interracial pornographic films, which most commonly refers to black male/white female films, have increased in popularity, becoming one of the fastest-growing and biggest-selling genres. Interracial films that include black men and white women together have a majority audience of white male viewers.

According to Renée Christine Romano, "Alvin Poussaint, a black psychiatrist who studied Freedom Summer volunteers, theorized that black men in the movement saw white women more as symbolic conquests than as individuals. The black men he studied viewed "sexual intimacy with the white girls as a weapon of revenge against white society," Poussaint argued."

Asia

Inter-ethnic marriage in Southeast Asia dates back to the spread of Indian culture, Hinduism and Buddhism to the region. From the 1st century onwards, mostly male traders and merchants from the Indian subcontinent frequently intermarried with the local female populations in Cambodia, Burma, Champa, central Siam, the Malay Peninsula, and Malay Archipelago. Many Indianized kingdoms arose in Southeast Asia during the Middle Ages.

From the 9th century onwards, a large number of mostly male Arab traders from the Middle East settled down in the Malay Peninsula and Malay Archipelago, and they intermarried with the local Malay, Indonesian and Filipina female populations. This contributed to the spread of Islam in Southeast Asia. From the 14th to the 17th centuries, many Chinese, Indian and Arab traders settled down within the maritime kingdoms of Southeast Asia and intermarried with the local female populations. This tradition continued among Portuguese traders who also intermarried with the local populations. In the 16th and 17th centuries, thousands of Japanese people also travelled to Southeast Asia and intermarried with the local women there.

The white men who sailed to Asia in the 16th century left a generation of mixed-race offspring that, at the high point of empire building, was more than one-million strong. Some 100,000 Amerasians stayed in Vietnam after the fall of Saigon. During and after the Indonesian National Revolution (1945–1965) around 300,000 people, pre-dominantly Eurasians, left Indonesia to go to the Netherlands.

In the 19th century and early 20th century, there was a network of small numbers of Chinese prostitutes being trafficked across Asia, in countries such as China, Japan, Korea, Singapore and British India, in what was then known as the 'Yellow Slave Traffic'. There was also a network of prostitutes from continental Europe being trafficked to India, Ceylon, Singapore, China and Japan at around the same time, in what was then known as the 'White Slave Traffic'.

During World War II, Japanese soldiers engaged in war rape during their invasions across East Asia and Southeast Asia. The term "comfort women" is a euphemism for the estimated 200,000, mostly Korean and Chinese, women who were forced into prostitution in Japanese military brothels during World War II. Some Dutch women, captured in Dutch colonies in Asia, were also forced into sexual slavery.

Sex tourism has emerged in the late 20th century as a controversial aspect of Western tourism and globalization. Sex tourism is typically undertaken internationally by tourists from wealthier countries. Author Nils Ringdal alleged that three out of four men between the ages of 20 and 50 who have visited Asia or Africa have paid for sex.

Central Asia

Today central asians are a mixed race of various peoples such as Mongols, Turkics, Iranics. The Mongol invasion of Central Asia in 13th century resulted in the massacre of mostly Iranic population and other Indo-European people as well as large numbers of intermarriage and assimilation. Modern Genetic shows that Central Asian Turkic people and Hazara are an mixture of Northeast asians and Indo-european people. Caucasian ancestry is prevelant in almost all central asian turkic people. Kazakhs, Hazara, Karakalpaks have more European mtdna than European y-dna, Kyrgyz have mostly European y-dna with substanial European mtdna. Other Turkic people like Uyghurs, Uzbeks, have mostly European y-dna but also an significant high percentages of European mtdna. Turkmen have predominately European y-dna and mtdna.

China

There have been various periods in the history of China where large numbers of Arabs, Persians and Turks from the "Western Regions" (Central Asia and West Asia) migrated to China, beginning with the arrival of Islam during the Tang Dynasty in the 7th century. Due to the majority of these immigrants being male, they often intermarried with local Chinese females. While intermarriage was initially discouraged by the Tang Dynasty, it was later encouraged during the Song Dynasty, which allowed third-generation immigrants with official titles to intermarry with Chinese imperial princesses. Immigration to China increased under the Mongol Empire, when large numbers of West and Central Asians were brought over to help govern Yuan China in the 13th century.

Iranian, Arab, and Turkic women also migrated to China and mixed with Chinese. Iranian women as dancers were in demand in China during this period. During the Sui dynasty, ten young dancing girls were sent from Persia to China. During the Tang dynasty bars were often attended by Iranian or Sogdian waitresses who performed dances for clients.

During the Five Dynasties and Ten Kingdoms Period(Wudai) (907-960), there are examples of Persian women marrying Chinese emperors. Some Chinese officials from the Song Dynasty era also married women from Dashi (Arabia).

Of the Han Chinese Li family in Quanzhou, Li Nu, the son of Li Lu, visited Hormuz in Persia in 1376, married a Persian or an Arab girl, and brought her back to Quanzhou. He then converted to Islam, Li Nu was the ancestor of the Ming Dynasty reformer Li Chih.

By the 14th century, the total population of Muslims in China had grown to 4 million. After Mongol rule had been overthrown by the Ming Dynasty in 1368, this led to a violent Chinese backlash against West and Central Asians. In order to contain the violence, the Ming administration instituted a policy where all West and Central Asian males were required to intermarry with native Chinese females, hence assimilating them into the local population. Their descendants are today known as the Hui people. Caucasian ancestry is also prevalent among the Hui people, 6.7% Hui people's maternal genetics have an Caucasian origin, while slightly over 30% paternal genetics also have an Caucasian origin. Han women who married Hui men became Hui, and Han men who married Hui women also became Hui.

In the frontier districts of Sichuan, numerous half Chinese-Tibetans were found. Tibetan women were glad to marry Chinese traders and soldiers. Some Chinese traders married Tibetan girls. Traders and officials in ancient times were often forbidden to bring Chinese women with them to Tibet, so they had to marry Tibetan women, the male offspring where considered Chinese and female offspring as Tibetan. Special names were used for these children of Chinese fathers and Tibetan mothers. They are often assimilated into the Tibetan population. Chinese and Nepalese in Tibet married Tibetan women.

In Qinghai, premarital sex between Tibetan girls and Han Chinese was common, some Tibetan girls boasted of their sexual conquests of Han Chinese boys. Dolkar, a Tibetan girl working as a guard at Drapchi Prison married a Chinese.

Chinese men also married Turkic Uyghur women in Xinjiang from 1880–1949. Sometimes poverty influenced Uyghur women to marry Chinese. These marriages were not recognized by local mullahs since Muslims women were not allowed to marry non Muslim men under Islamic law.

This did not stop the women because they enjoyed advantages, not being subject to Islamic law and they were not subjected to certain taxes. Uyghur women married to Chinese also did not have to wear a veil and they received their husband's property upon his death. These women were forbidden from having burial in Muslim graves. The children of Chinese men and Uyghur women were considered as Uyghur. Some Chinese soldiers had Uyghur women as temporary wives, and after the man's military service was up, the wife was left behind or sold, and if it was possible, sons were taken, and daughters were sold.

European travellers noted that many Han Chinese in Xinjiang married Uyghur (who were called turki) women and had children with them. A Chinese was spotted with a "young" and "good looking" Uyghur wife and another Chinese left behind his Uyghur wife and child in Khotan.

After 1950, some intermarriage between Han and Uyghur continued. A Han married a Uyghur woman in 1966 and had three daughters with her, and other cases on intermarriage also continued.

Hong Kong

South Asians have been living in Hong Kong throughout the colonial period, before the partition of India into the nations of India and Pakistan. They migrated to Hong Kong and worked as police officers as well as army officers during colonial rule. 25,000 of the Muslims in Hong Kong trace their roots back to what is now Pakistan. Around half of them belong to 'local boy' families, Muslims of mixed Chinese and South Asian ancestry, descended from early Indian/Pakistani Muslim immigrants who took local Chinese wives and brought their children up as Muslims.

Indian Subcontinent

The Indian subcontinent has a long history of inter-ethnic marriage dating back to ancient history. Various groups of people have been intermarrying for millennia in South Asia, including groups as diverse as the Dravidian, Indo-Aryan, Austro-Asiatic and Tibeto-Burman peoples. Invading Macedonians, Greeks, Scythians, Huns, Persians, Mongols (known as Mughals), and Europeans took Indian wives.

3,000–8,000 years ago, Indo-European-speaking nomadic groups from Europe, the Near East, Anatolia, and the Caucasus migrated to India. According to 19th-century British historians, it was these "Aryans" who established the caste system, an elitist form of social organization that separated the "light-skinned" Indo-Aryan conquerors from the "conquered dark-skinned" indigenous Dravidian population through enforcement of "racial endogamy". Much of this was simply conjecture, fueled by British imperialism; British policies of divide and rule as well as enumeration of the population into rigid categories during the tenure of British rule in India contributed towards the hardening of these segregated caste identities. Since the independence of India from British rule, the British fantasy of an "Aryan Invasion and subjugation of the dark skinned Dravidians in India" has become a staple polemic in South Asian geopolitics, including the propaganda of Indophobia in Pakistan. There is no decisive theory as to the origins of the caste system in India, and globally renowned historians and archaeologists like Jim Shaffer, J.P. Mallory, Edwin Bryant, and others, have disputed the claim of "Aryan Invasion".

Some researchers claim that genetic similarities to Europeans were more common in members of the higher ranks. Their findings, published in Genome Research, supported the idea that members of higher castes are more closely related to Europeans than are the lower castes. According to the research, invading European populations were predominantly male who intermarried with local females and formed the upper castes i.e. the local females had upward mobility in caste which was denied to local males. However, other researchers have criticized and contradicted this claim. A study by Joanna L. Mountain et al. of Stanford University concluded that there was "no clear separation into three genetically distinct groups along caste lines", although "an inferred tree revealed some clustering according to caste affiliation". A 2006 study by Ismail Thanseem et al. of Centre for Cellular and Molecular Biology (India) concluded that the "lower caste groups might have originated with the hierarchical divisions that arose within the tribal groups with the spread of Neolithic agriculturalists, much earlier than the arrival of Aryan speakers", and "the Indo-Europeans established themselves as upper castes among this already developed caste-like class structure within the tribes." A 2006 genetic study by the National Institute of Biologicals in India, testing a sample of men from 32 tribal and 45 caste groups, concluded that the Indians have acquired very few genes from Indo-European speakers. More recent studies have also debunked the claims that so-called "Aryans" and "Dravidians" have a "racial divide". A study conducted by the Centre for Cellular and Molecular Biology in 2009 (in collaboration with Harvard Medical School, Harvard School of Public Health and the Broad Institute of Harvard and MIT) analyzed half a million genetic markers across the genomes of 132 individuals from 25 ethnic groups from 13 states in India across multiple caste groups. The study establishes, based on the impossibility of identifying any genetic indicators across caste lines, that castes in South Asia grew out of traditional tribal organizations during the formation of Indian society, and were not the product of any Aryan invasion and subjugation of Dravidian people.

Rape and enslavement of Hindu women by invading Islamic armies was very common. For example, according to the Chachnama by Qazi Ismail, during the Arab invasion of Sindh (712 CE), Muhammad bin Qasim first attacked and conquered Debal. It was situated on the sea-coast (not far from modern Karachi). All males of the age of seventeen and upwards were put to the sword and their women and children were enslaved. 700 Hindu women, who were under the protection of Budh (that is, had taken shelter in the temple), were all captured with their valuable ornaments, and adorned with jewels. Muhammad despatched one-fifth of the legal spoil to Hajjaj which included seventy-five damsels, while the other four-fifths were distributed among the soldiers. Thereafter whichever places he attacked like Rawar, Sehwan, Dhalila, Brahmanabad and Multan, Hindu soldiers and men with arms were slain, many women of the higher classes immolated themselves in Jauhar. Most others became prize of the victors. These women were enslaved and converted, and batches of them were despatched to the Caliph in regular installments. For example, after Rawar was taken, many women and children were enslaved, of whom 30 were young ladies of royal blood. Muhammad Qasim sent all these to Hajjaj, who forwarded them to Walid the Khalifa. He sold some of these female slaves of royal birth, and some he presented to others. In Northern India, female slaves captured after every campaign of the marching army, were converted and married to Muslim soldiers who settled down in colonies established in places like Mansura, Kuzdar, Mahfuza and Multan. During the Islamic involvement in India, it was normal for kings to possess harems filled with native Hindu women won as booties of war. The most famous one was of Akbar's harem, which had over 5000 women. Most of these Muslim soldiers were Turks, Afghans, Persians, Mongols (Mughals), other Central Asians and Arabs.

There are even cases of Indian princesses marrying kings abroad. For example, the Korean text Samguk Yusa about the Gaya kingdom (it was absorbed by the kingdom of Silla later), indicates that in 48 AD, King Kim Suro of Gaya (the progenitor of the Gimhae Kim clan) took a princess (Princess Heo) from the "Ayuta nation" (which is the Korean name for the city of Ayodhya in North India) as his bride and queen. Princess Heo belonged to the Mishra royal family of Ayodhya. According to the Samguk Yusa, the princess had a dream about a heavenly fair handsome king from a faraway land who was awaiting heaven's anointed ride. After Princess Heo had the dream, she asked her parents, the king and queen of Ayodhya, for permission to set out and seek the foreign prince, which the king and queen urged with the belief that god orchestrated the whole fate. That king was no other than King Kim Suro of the Korean Gaya kingdom.

In Goa, a Portuguese colony in India, during the late 16th century and 17th century, there was a community of over thousand Japanese slaves and traders, who were either Japanese Christians fleeing persecution in Japan, or young Japanese women and girls brought or captured as sexual slaves by Portuguese traders and their South Asian lascar crew members from Japan. In both cases, they often intermarried with the local population in Goa.

Interracial marriages between European men and Indian women were very common during colonial times. Most of these Indian women usually were Muslim belonging to aristocratic families and families with royal ancestry.

According to the historian William Dalrymple, about one in three European men had Indian wives in colonial India. This was primarily because the Europeans (mostly Dutch, British, French and Portuguese and up to a lesser extent Swedes and Danes) came to India when they were young and there were very few white women available in India. The most famous of such interracial liaisons was between the Hyderabadi noblewoman Khair-un-Nissa and the Scottish resident James Achilles Kirkpatrick.

The 600,000-strong Anglo-Indian community was formed by British soldiers taking Indian women as wives. Such relationships have had a strong influence on the arts. Lakmé, an opera by the Frenchman Léo Delibes, deals with the romantic relationship between the British officer Gérald and the daughter of a Hindu high priest Lakmé (Laxmi in Sanskrit).

In Ceylon (present day Sri Lanka), interracial relationships between Dutch, British and Portuguese men and local women were common. The 65,000-strong Burgher community was formed by the interracial marriages of Dutch and Portuguese men with local Sinhalese and Tamil women. In addition to intermarriage, inter-ethnic prostitution in India was also fairly common at the time, when British officers would frequently visit Indian nautch dancers. In the mid-19th century, there were around 40,000 British soldiers but fewer than 2,000 British officials present in India. Many British and other European officers had their own harems made up of Indian women similar to those the Nawabs and kings of India had. In the 19th century and early 20th century, thousands of women and girls from continental Europe were also trafficked into British India (and Ceylon), where they worked as prostitutes servicing both British soldiers and local Indian (and Ceylonese) men.

As British females began arriving in British India in large numbers from the early-to-mid-19th century, miscegenation became increasingly uncommon in India as British women increasingly disapproved of relationships between local woman and their men. Interracial relationships were also despised after the events of the Sepoy Mutiny of 1857, also known as “India's First War of Independence”, where some Indian sepoys rebelled against the British East India Company.

The idea of protecting British female chastity from the “lustful Indian male” had a significant influence on the policies of the British Raj in order to prevent racial miscegenation between the British females and the native Indian male population. While some restrictive policies were imposed on British females in order to protect them from miscegenation, most of these policies were directed against native Indian males.

For example, the 1883 Ilbert Bill, which would have granted Indian judges the right to judge British offenders, was opposed by many British colonialists on the grounds that Indian judges cannot be trusted in dealing with cases involving British females. In the aftermath of the 1919 Amritsar Massacre, the long-held stereotype of Indian males as dark-skinned rapists lusting after white British females was challenged by several novels such as A Passage to India (1924) and The Jewel in the Crown (1966), both of which involve an Indian male being wrongly accused of raping a British female.

When Burma was ruled under the administration of British India, millions of Indians, mostly Muslim, migrated there. The small population mixed descendants of Indian males and local Burmese females are called "Zerbadees", often in a pejorative sense implying mixed race.

Japan

Inter-ethnic marriage in Japan dates back to the 7th century, when Chinese and Korean immigrants began intermarrying with the local Japanese population. In the 1590s, over 50,000 Koreans were forcibly brought to Japan during Hideyoshi's invasions of Korea, where they intermarried with the local population. In the 16th and 17th centuries, around 58,000 Japanese travelled abroad, many of whom intermarried with the local women in Southeast Asia. During the anti-Christian persecutions in 1596, many Japanese Christians fled to Macau and other Portuguese colonies such as Goa, where there was a community of Japanese slaves and traders by the early 17th century. Intermarriage with the local populations in these Portuguese colonies also took place. Portuguese traders in Japan also intermarried with the local Christian women.

From the 15th century, Chinese, Korean and other Far Eastern visitors frequented brothels in Japan. This practice later continued among visitors from the "Western Regions", mainly European traders. This began with the arrival of Portuguese ships to Japan in the 16th century. Portuguese visitors and their South Asian (and sometimes African) crewmembers often engaged in slavery in Japan, where they brought Japanese slaves who were taken to Macau and other Portuguese colonies in Southeast Asia, the Americas, and India. Later European East India companies, including those of the Dutch and British, also engaged in prostitution in Japan.

In the early part of the Shôwa era, Japanese governments executed a eugenic policy to limit the birth of children with inferior traits, as well as aiming to protect the life and health of mothers. Family Center staff also attempted to discourage marriage between Japanese women and Korean men who had been recruited from the peninsula as laborers following its annexation by Japan in 1910. In 1942, a survey report argued that "the Korean laborers brought to Japan, where they have established permanent residency, are of the lower classes and therefore of inferior constitution...By fathering children with Japanese women, these men could lower the caliber of the Yamato minzoku." In 1928, journalist Shigenori Ikeda promoted the 21 December as the blood-purity day (junketsu de) and sponsored free blood-test at the Tokyo Hygiene laboratory. By the early 1930s, detailed "eugenic marriage" questionnaires were printed or inserted in popular magazines for public consumption. Promoters like Ikeda were convinced that these marriage surveys would not only insure the eugenic fitness of spouses but also help avoid class differences that could disrupt and even destroy marriage. The goal was to create a database of individuals and their entire households which would enable eugenicists to conduct in-depth surveys of any given family's genealogy.

To prevent venereal diseases and rape by Japanese soldiers and to provide comfort to soldiers and head off espionage, the Imperial Japanese Army established "comfort stations" in the Greater East Asia Co-Prosperity Sphere where around 200,000 women, mostly from Korea and China, were recruited or kidnapped by the Kempeitai or the Tokeitai as comfort women.

One of the last eugenic measures of the Shôwa regime was taken by the Higashikuni government. On 19 August 1945, the Home Ministry ordered local government offices to establish a prostitution service for Allied soldiers to preserve the "purity" of the "Japanese race". The official declaration stated that : "Through the sacrifice of thousands of "Okichis" of the Shôwa era, we shall construct a dike to hold back the mad frenzy of the occupation troops and cultivate and preserve the purity of our race long into the future...."

According to Peter Schrijvers in "The GI War against Japan: American Soldiers in Asia and the Pacific during World War II", rape "reflects a burning need to establish total dominance of the other" the enemy. According to Xavier Guillaume, US soldiers' rape of Japanese women was "general practice". Schrijvers states regarding rapes on Okinawa that "The estimate of one Okinawan historian for the entire three-month period of the campaign exceeds 10,000. A figure that does not seem unlikely when one realizes that during the first 10 days of the occupation of Japan there were 1,336 reported cases of rape of Japanese women by American soldiers in Kanagawa prefecture alone".

However, despite being told by the Japanese military that they would suffer rape, torture and murder at the hands of the Americans, Japanese civilians "were often surprised at the comparatively humane treatment they received from the American enemy." According to Islands of Discontent: Okinawan Responses to Japanese and American Power by Mark Selden, the Americans "did not pursue a policy of torture, rape, and murder of civilians as Japanese military officials had warned." Japanese society, with its ideology of homogeneity, has traditionally been intolerant of ethnic and other differences.

Men or women of mixed ancestry, foreigners, and members of minority groups faced discrimination in a variety of forms. In 2005, a United Nations report expressed concerns about racism in Japan and that government recognition of the depth of the problem was not total. In 2005, Japanese Minister Taro Aso called Japan a "one race" nation.

Korea

Inter-ethnic marriage in Korea dates back to the arrival of Muslims in Korea during the Middle Ages, when Persian and Turkic navigators, traders and slaves settled in Korea and married local Korean people. Some assimilation into Buddhism and Shamanism eventually took place, owing to Korea's geographical isolation from the Muslim world. There are several Korean clans that are descended from such intermarriages. For example, the Deoksu Jang clan, claiming some 30,000 Korean members, view Jang Sunnyong, a Central Asian who married a Korean female, as their ancestor. Another clan, Gyeongju Seol, claiming at least 2,000 members in Korea, view a Central Asian (probably an Uyghur) named Seol Son as their ancestor.

There are even cases of Korean kings marrying princesses from abroad. For example, the Korean text Samguk Yusa about the Gaya kingdom (it was absorbed by the kingdom of Silla later), indicate that in 48 AD, King Kim Suro of Gaya (the progenitor of the Gimhae Kim clan) took a princess (Princess Heo) from the "Ayuta nation" (which is the Korean name for the city of Ayodhya in North India) as his bride and queen. Princess Heo belonged to the Mishra royal family of Ayodhya. According to the Samguk Yusa, the princess had a dream about a heavenly fair handsome king from a far away land who was awaiting heaven's anointed ride. After Princess Heo had the dream, she asked her parents, the king and queen of Ayodhya, for permission to set out and seek the foreign prince, which the king and queen urged with the belief that god orchestrated the whole fate. That king was no other than King Kim Suro of the Korean Gaya kingdom.

6,423 Korean women married US military personnel as war brides during and immediately after the Korean War. The average number of Korean women marrying US military personnel each year was about 1,500 per year in the 1960s and 2,300 per year in the 1970s.

International marriages now make up 13% of all marriages in South Korea. Most of these marriages are unions between a Korean male and a foreign female usually from China, Japan, Vietnam, the Philippines, United States, Mongolia, Thailand, and Russia. On the other hand, Korean females have married foreign males from Japan, China, the United States, Bangladesh, Pakistan, Philippines, and Nepal. Between 1990 and 2005, there have been 159,942 Korean males and 80,813 Korean females married to foreigners.

South Korea is among the world's most ethnically homogeneous nations. Koreans have traditionally valued an unmixed blood as the most important feature of Korean identity. The term "Kosian", referring to someone who has a Korean father and a non-Korean mother, is considered offensive by some who prefer to identify themselves or their children as Korean. Moreover, the Korean office of Amnesty International has claimed that the word "Kosian" represents racial discrimination. Kosian children, like those of other mixed-race backgrounds in Korea, often face discrimination. There are an estimated 35,000 mixed-raced South Koreans, most of them half Caucasian, according to the Pearl Buck Foundation. Discrimination is far worse against those who have African American fathers.

Malaysia and Singapore

In West Malaysia and Singapore, the majority of inter-ethnic marriages are between Chinese and Indians. The offspring of such marriages are informally known as "Chindian", though the Malaysian government only classifies them by their father's ethnicity. As the majority of these intermarriages usually involve an Indian groom and Chinese bride, the majority of Chindians in Malaysia are usually classified as "Indian" by the Malaysian government. As for the Malays, who are predominantly Muslim, legal restrictions in Malaysia make it uncommon for them to intermarry with either the Indians, who are predominantly Hindu, or the Chinese, who are predominantly Buddhist and Taoist. Non-Muslims are required to convert to Islam in order to marry Muslims. However, this has not entirely stopped intermarriage between the Malays and the Chinese and Indians. There are more Indian-Malay marriages than there are Chinese-Malay ones; this has led to the emergence of a large Muslim Indian society that is most famously known for their 24-hour coffee shops and active involvement in the textile industry. The Muslim Chinese community, on the other hand, is small and has only a negligible impact on the socio-economy and demography of the region.

It is common for Arabs in Singapore and Malaysia to take local Malay wives, due to a common Islamic faith. The Chitty people, in Singapore and the Malacca state of Malaysia, are a Tamil people with considerable Malay descent, which was due to the first Tamil settlers taking local wives, since they did not bring along any of their own women with them. According to government statistics, the population of Singapore as of September 2007 was 4.68 million, of whom multiracial people, including Chindians and Eurasians, formed 2.4%.

In the East Malaysian states of Sabah and Sarawak, there have been many incidents of intermarriage between Chinese and native tribes such as the Murut and Dusun in Sabah, and the Iban and Bisaya in Sarawak. This phenomenon has resulted in a potpourri of cultures in both states where many people claiming to be of native descent have some Chinese blood in them, and many Chinese have native blood in them. The offspring of these mixed marriages are called 'Sino-(name of tribe)', e.g. Sino-Dusun. Normally, if the father is Chinese, the offspring will adopt Chinese culture and if the father is native then native culture will be adopted, but this is not always the case. These Sino-natives are usually fluent in Malay and English. A smaller number are able to speak Chinese dialects and Mandarin, especially those who have received education in vernacular Chinese schools.

Burma

Burmese Muslims are the descendants of Bengalis, Indian Muslims, Arabs, Persians, Turks, Pathans, Chinese Muslims and Malays who settled and intermarried with the local Burmese population and other Burmese ethnic groups such as the Rakhine, Shan, Karen, and Mon.

The oldest Muslim group in Burma (Myanmar) are the Rohingya people, who some believed are descended from Bengalis who intermarried with the native females in the Rakhine State after the 7th century, but this is just an theory. When Burma was ruled by the British India administration, millions of Indians, mostly Muslim, migrated there. The small population of mixed descendants of Indian males and local Burmese females are called "Zerbadees", often in a pejorative sense implying mixed race. The Panthays, a group of Chinese Muslims descended from West Asians and Central Asians, migrated from China and also intermarried with local Burmese females.

In addition, Burma has an estimated 52,000 Anglo-Burmese people, descended from British and Burmese people. Anglo-Burmese people frequently intermarried with Anglo-Indian immigrants, who eventually assimilated into the Anglo-Burmese community.

Philippines

Historically, admixture has been an ever present and pervading phenomenon in the Philippines. The Philippines were originally settled by Australoid peoples called Negritos (different from other australoid groups) which now form the country's aboriginal community. Some Admixture may have occurred between this earlier group and the mainstream Malayo-Polynesian population.

A considerable number of the population in the town of Cainta, Rizal, are descended from Indian soldiers who mutinied against the British Indian Army when the British briefly occupied the Philippines in 1762 to 1763. These Indian soldiers called Sepoy settled in town and intermarried with native females. The Sepoy ancestry of Cainta is very visible today, particularly in Barrio

Dayap near Brgy. Sto Nino. Their unique physical characteristics make them distinct from the average Filipinos. There has been a Chinese presence in the Philippines since the 9th century. However, large-scale migrations of Chinese to the Philippines only started during the Spanish colonial era, when the world market was opened to the Philippines. It is estimated that among Filipinos, 10%–20% have some Chinese ancestry and 1.5% are "full-blooded" Chinese.

According to the American anthropologist Dr. H. Otley Beyer, the ancestry of Filipinos is 2% Arab. This dates back to when Arab traders intermarried with the local Malay Filipina female populations during the pre-Spanish history of the Philippines. Major Arab migration to the Philippines coincided with the spread of Islam in the region. Filipino-Muslim royal families from the Sultanate of Sulu and the Sultanate of Maguindanao claim Arab descent even going as far as claiming direct lineage from the prophet Mohammad. Such intermarriage mostly took place around the Mindanao island area, but the arrival of Spanish Conquistadors to the Philippines abruptly halted the spread of Islam further north into the Philippines. Intermarriage with Spanish people later became more prevalent after the Philippines was colonized by the Spanish Empire.

When the Spanish colonized the Philippines, a significant portion of the Filipino population mixed with the Spanish. When the United States took the Philippines from Spain during the Spanish-American War, much intermixing of Americans, both white and black, took place on the island of Luzon where the USA had a Naval Base and Air Force Base even after the USA gave the Philippines independence after World War II. First children and descendants of male Filipino population with Spanish surnames who intermarried with white American female population may be considered Spanish mestizos. The descendants of Filipinos and Europeans are today known as mestizos, following the term used in other former Spanish colonies.

Much mixing with the Japanese also took place due to the war rapes of Filipina women during World War II. Today there is an increasing number of Japanese men marrying Filipina woman and fathering children by them whose family remain behind in the Philippines and are financially supported by their Japanese fathers who make regular visits to the Philippines. Today mixed-race marriages have a mixed reaction in the Philippines, most urban centers like Manila and Cebu are more willing to accept interracial marriages than rural areas, plus there is more approval if the Filipina marries out than a Filipino male.

Europe, Germany

During the tempestuous years following World War I, the French Army occupied the Rhineland, utilising African soldiers amongst their forces. Their children were known as "Rhineland Bastards".

Beginning in 1933, the mainstream Nazi anti-Semitism considered the Jews as being a group of people bound by close, so-called genetic (blood) ties, to form a unit, which one could not join or secede from. The influence of Jews had been declared to have a detrimental impact on Germany, in order to justify the discriminations and persecutions of Jews. To be spared from those, one had to prove one's affiliation with the group of the Aryan race, as conceived by the Nazis.

It was paradoxical that neither genetic tests nor allegedly racial outward features in one's physiognomy determined one's affiliation, although the Nazis talked a lot about physiognomy, but only the records of the religious affiliations of one's grandparents decided it. However, while earlier the grandparents had still been able to choose their religion, their grandchildren in the Nazi era were compulsorily categorised as Jews, thus non-Aryans, if three or four grandparents had been enrolled as members of a Jewish congregation, regardless of whether the persecuted themselves were Jews according to the Halachah (roughly meaning: Jewish by birth from a Jewish mother or by conversion), apostates, irreligionists or Christians.

The Nuremberg Laws of 1935 forbade persons racially regarded as so-called Aryans and non-Aryans to marry; this included all marriages where at least one partner was a German citizen. Non-Aryans comprised mostly Jewish Germans and Gentile Germans of Jewish descent. However, Germans of extra-European and especially of African descent and Germans regarded as belonging to the minority group of Sinti and Roma were also considered as non-Aryans.

Eventually children – whenever born – within a mixed marriage, as well as children from extramarital mixed relationships born until July 31, 1936, were discriminated against as Mischlinge. However, children later born to mixed parents, not yet married as at the passing of the Nuremberg Laws, were to be discriminated against as Geltungsjuden, regardless of whether the parents had meanwhile married abroad or remained unmarried. Eventually children who were enrolled in a Jewish congregation were also subject to discrimination as Geltungsjuden.

Geltungsjuden were subjected to varying degrees of forced labour in 1940, partly ordered for all Jewish-classified spouses, either only for Jewish-classified husbands or only exempting Jewish-classified wives taking care of minor children. No documents indicate the exemption of a mixed marriage and especially of its Jewish-classified spouse from some persecutions.

Systematic deportations of Jewish Germans and Gentile Germans of Jewish descent started on October 18, 1941. German Jews and German Gentiles of Jewish descent living in mixed marriages were in fact mostly spared from deportation. In the event that a mixed marriage ended by the death of the so-called Aryan spouse or the divorce of the Jewish-classified spouse, the Jewish-classified spouse residing within Germany was usually deported soon after unless the couple still had minor children not counted as Geltungsjuden. In March 1943 an attempt to deport the Berlin-based Jews and Gentiles of Jewish descent, living in non-privileged mixed marriages, failed due to public protest by their in-laws of so-called Aryan kinship (Rosenstraße protest). Also the Aryan-classified husbands and Mischling-classified children (starting at the age of 16) from mixed marriages were taken by the Organisation Todt for forced labour, starting in autumn 1944.

A last attempt, undertaken in February/March 1945, ended because the extermination camps were already liberated. However, 2,600 from all over the Reich were deported to Theresienstadt, of whom most survived the last months until their liberation.

With the defeat of Nazi Germany in 1945 the laws banning so-called mixed marriages were lifted again. If couples who had already lived together during the Nazi era had remained unmarried due to the legal restrictions then got married after the war, their date of marriage was legally retroactively backdated if they wished it to the date they formed a couple. Even if one spouse was already dead, the marriage could be retroactively recognised. In the West German Federal Republic of Germany 1,823 couples applied for recognition, which was granted in 1,255 cases.

It is estimated that up to 7,000 postwar black German children with black GI fathers and white German mothers were adopted by Americans.

Hungary

The Avars, Asiatic nomads who during the late 6th and 7th centuries had formed an extensive empire largely inhabited by conquered Slavs, made the agricultural Slavs pay taxes, and used their wives and daughters as concubines.

The Hungarians are thought to have originated in an ancient Finno-Ugric population that originally inhabited the forested area between the Volga River and the Ural Mountains. At the time of the Magyar migration in the 10th century, the present-day Hungary was inhabited by Slavs, numbering about 200,000, who were either assimilated or enslaved by the Magyars.

During the Russian campaign in the 13th century, the Mongols drove some 40,000 Cuman families, a nomadic tribe, west of the Carpathian Mountains. The Iranian Jassic people came to Hungary together with the Cumans after they were defeated by the Mongols. Over the centuries they were fully assimilated into the Hungarian population. Rogerius, a monk who survived the Mongol invasion of Hungary, pointed out not only the genocidal element of the occupation, but also that the Mongols especially "found pleasure" in humiliating women.

Iberian Peninsula

In ancient history, the Iberian Peninsula was frequently invaded by foreigners who intermarried with the native population. One of the earliest foreign groups to arrive in the region were the Indo-European Celts who intermarried with the pre-Indo-European Iberians in prehistoric Iberia. They were later followed by the Semitic Phoenicians and Carthaginians and the Indo-European Romans who intermarried with the pre-Roman peoples of the Iberian Peninsula during Classical Antiquity. They were in turn followed by the Germanic Visigoths, Suebi and Vandals and the Iranic Sarmatians and Alans who also intermarried with the local population in Hispania during late Antiquity.

In the 6th century, the region was reconquered by the Byzantine Empire (Eastern Roman Empire), when Byzantine Greeks also settled there, before the region was lost again to the Visigothic Kingdom less than a century later. After the Umayyad conquest of Hispania in the 8th century, the Islamic state of Al-Andalus was established in the Iberian Peninsula. Islamic marital law is by itself an anti-miscegenation law, since it allows a Muslim male to marry Christian and Jewish females but not a Muslim female to marry a Christian or Jewish male.

The offspring of such marriages were known as Muladi or Muwallad, an Arabic term still used in the modern Arab world to refer to people with Arab fathers and non-Arab mothers. Some sources consider this term the origin for the Spanish word Mulato. However, the Real Academia Española does not endorse such etymology. In addition, many Muladi were also descended from Saqaliba (Slavic) slaves taken from Eastern Europe via the Arab slave trade.

After the Reconquista, which was completed in 1492, most of the Moors were forced to either flee to Morocco or convert to Christianity. The ones who converted to Christianity were known as Moriscoes, and they were often persecuted by the Spanish Inquisition as suspects of heresy on the basis of the Limpieza de sangre ("Cleanliness of blood") doctrine, under which anti-miscegenation laws were implemented in the Iberian Peninsula.

Anyone whose ancestors had miscegenated with the Moors or Jews were suspicious of secretly practicing Islam or Judaism, so were often particularly monitored by the Inquisition. The claim to universal hidalguía (lowest nobility) of the Basques was justified by erudites like Manuel de Larramendi (1690–1766) because the Arab invasion had not reached the Basque territories, so it was believed that Basques had maintained their original purity, while the rest of Spain was suspect of miscegenation. In fact, the Arab invasion also reached the Basque country and there had been a significant Jewish minority in Navarre, but the hidalguía helped many Basques to official positions in the administration. In December 2008, an important genetic study revealed that the religious conversions of Jews and Muslims have had a profound impact on the population of the Iberian Peninsula. This study indicated a Sephardic Jewish mean admixture of about 20% and a North African admixture of about 11%.

Italian Peninsula

As was the case in other regions conquered by Muslims, it was acceptable in Islamic marital law for a Muslim male to marry Christian and Jewish females in southern Italy when under Islamic rule between the 8th and 11th centuries. In this case, most intermarriages were between Arab and Berber males from North Africa and the local Greek, Roman and Italian females of Sicily and southern Italy.

Such intermarriages were particularly common in the Emirate of Sicily, where one writer visiting the place in the 970s expressed shock at how common it was in rural areas. After the Norman conquest of southern Italy, all Muslim citizens (whether foreign, native or mixed) of the Kingdom of Sicily were known as "Moors". After a brief period of Arab-Norman culture had flourished under the reign of Roger II of Sicily, later rulers had forced the Moors to either convert to Christianity or be expelled from the kingdom.

In Malta, Arabs and Italians from neighbouring Sicily and Calabria intermarried with the local inhabitants, who were descended from Phoenicians, Greeks, Romans and Vandals. The Maltese people are descended from such unions, and the Maltese language is descended from Siculo-Arabic.

In the Republic of Venice in northern Italy, it was common for foreign Arab and Berber traders, known to Europeans as the "Moors", to take local Italian wives. This became a subject matter in several William Shakespeare plays, most notably Othello, involving an inter-ethnic relationship between a Moorish Othello and his Venetian wife Desdemona, based on Giovanni Battista Giraldi's "Un Capitano Moro" which was itself inspired by an actual incident that occurred in Venice around 1508.

At times, the Italian city-states also played an active role in the Arab slave trade, where Moorish and Italian traders occasionally exchanged slaves. Leonardo da Vinci's mother Caterina, for example, was most likely a slave from the Middle East.

During World War II, France's Moroccan troops known as Goumiers committed war rapes in Italy after the Battle of Monte Cassino and in Germany. In Italy, victims of the mass rape committed after the Battle of Monte Cassino by Goumiers are known as Marocchinate. According to Italian sources, more than 7,000 Italian civilians, including women and children, were raped by Goumiers.

Southeastern and Eastern Europe

Vikings explored and eventually settled in territories in Slavic-dominated areas of Eastern Europe. By 950 AD these settlements were largely Slavicized through intermarriage with the local population. Eastern Europe was also an important source for the Arab slave trade at the time, when Saqaliba (Slavic) slaves were taken to the Arab World, where the women and girls often served in harems, some of whom married their Arab masters. When the Mongol Empire annexed much of Eastern Europe in the 13th century, the Mongols also intermarried with the local population and often engaged in war rape during the Mongol invasion of Europe.

In the 11th century, the Byzantine territory of Anatolia was conquered by the Seljuq Turks, who came from Turkestan in Central Asia. Their Ottoman Turkish descendants went on to annex the Balkans and much of Eastern Europe in the 15th and 16th centuries. Due to Islamic marital law allowing a Muslim male to marry Christian and Jewish females, it was common in the Ottoman Empire for Turkish males to intermarry with European females. For example, various sultans of the Ottoman Dynasty often had Greek (Rûm), Slavic (Saqaliba), Venetian, Northcaucasian and French wives.

Some of these European wives exerted great influence upon the empire as Valide Sultan ("Mother-Sultan"), some famous examples including Roxelana, a Slavic harem slave who later became Suleiman the Magnificent's favourite wife, and Aimée du Buc de Rivéry, wife of Abdul Hamid I and cousin of French Empress Josephine.

Due to the common occurrence of such intermarriages in the Ottoman Empire, they have had a significant impact on the ethnic makeup of the modern Turkish population in Turkey, which now differs from that of the Turkic population in Central Asia. In addition to intermarriage, the large harems of Ottoman sultans often consisted almost entirely of female concubines who were of Christian European origin.

The degree of miscegenation is very high in the former Soviet Union. Interethnic marriages made up about 22% of all marriages in Moscow, according to the figures from 1995. The majority of such marriages consist among Russian, Ukrainians, Byelorussians and other Slavic.

Annually, many Russian women end up as prostitutes in India, China, Japan, or South Korea. Bars in major Chinese cities now offer blond, blue-eyed Russian "hostesses", while in Tokyo, Russian girls are the latest addition to the menu in fancy "hostess" bars. Russian prostitutes are found everywhere in Macau. Their customers are almost exclusively wealthy Chinese, Japanese and Korean businessmen. Eastern European women also end up in Turkey and the United Arab Emirates.

Low levels of mixed ancestry are, in some areas (especially urban), almost universal, and generally go entirely ignored and unnoticed unless persons wish to identify themselves with ethnic minorities. Highly visible divergence from the local ethnic majority is also treated differently, depending on whether the individual identifies with the local culture or not. In modern times, attitudes towards miscegenation in the former Soviet Union vary greatly, depending on the race and gender of each partner.

For example, unions between Slavic males and Oriental (Asian) or Turkic, Northcaucasian, Southcaucasian and Finno-Ugric women are almost universally tolerated, and their children are generally identified and treated as members of the local ethnic majority.

However, unions between Slavic women and visibly non-Slavic men may meet varying degrees of discrimination, from light to none for Asian men (depending also on origin, whether they are immigrants or were born in the Soviet Union, and where in the Soviet Union they were born), to some hostility for Finno-Ugric, Turkic and Northcaucasian men (although much of this is due to the assumption of their faith as Muslim) and Jews, and quite high intolerance towards those who marry blacks or have children by them (young African-Russians in Moscow are often scornfully called 'Children of the Olympics', under the assumption that they were conceived by visiting tourists during the 1980 Moscow Olympic Games). The situation is also highly affected by self-identification, since many people of Asian or Turkic blood have assimilated to the point where they identify themselves as Russian/Ukrainian/etc. and are socially accepted as such.

United Kingdom

Even though 80 percent of the genetic characteristics of most indigenous Britons today have been passed down from a few thousand Ice Age hunters and the genetic makeup of today's indigenous population is much the same as it was 12,000 years ago, Britain has a fairly long history of inter-ethnic marriage among the various European populations that inhabited the island, including the Celtic, Roman, Anglo-Saxon and Anglo-Norman peoples. Although it is debateable whether these can be described as ethnic groups with distinct cultural identities.

The Celts cannot be described with a single ethnic identity and various tribal names were used and a 'Celtic' culture was shared in Europe despite the variety of languages. Norman and Vikings people rarely held onto an identity and allegiances were more to leaders and family rather than an ethnic group, there is no evidence the Vikings for example acted as an distinct ethnic group in Britain and so called Scandinavian markers may have been in Britain thousands of years in the Neolithic. Also there were Scandinavians and pre-

Viking Scandinavian influence amongst the Anglo-Saxons as shown in Beowulf and Sutton Hoo. There is no evidence the Anglo-Saxons used that term or indeed forced their languages or culture on local Britons, indeed there is no evidence the majority of local Britons used the term Celt or even spoke a Celtic language, indeed a Germanic language which evolved into English may have been spoke in large parts of Britain as Oppenheimer suggests.

The numbers of Vikings, Anglo-Saxons, Romans may be widely exaggerated, Viking, Anglo-Saxons armies may have been small. Intermarriage with non-European populations began in the late 15th century, with the arrival of the Romani people, who have Indian origins. The Romani in Britain intermarried with the local population and became known to the Romani as the Romanichal. In India, the British East India Company and other European soldiers intermarried with Indian women. The offspring of these mixed marriages between the British and Indians were known as Anglo-Indians. Indian wives sometimes accompanied their husbands back to Britain.

Inter-ethnic relationships have become increasingly accepted over the last several decades. As of 2001, 2% of all marriages in Britain are inter-ethnic. Despite having a much lower non-white population (9%), mixed marriages in the United Kingdom are as common as in the United States, although America has many fewer specific definitions of race (four racial definitions as opposed to the United Kingdom's 86). As of 2005, it is estimated that nearly half of British-born African-Caribbean males, a third of British-born African-Caribbean females, and a fifth of Indian and African males, have white partners. As of 2009, one in 10 children in the UK lives in a mixed-race family and two out of five Chinese women have partners of a different race.

Middle East

A Stanford team found the greatest diversity outside Africa among people living in the wide crescent of land stretching from the eastern shore of the Mediterranean to northern India. Not only was the region among the first colonized by the African migrants, they theorize, but the large number of European and East Asian genes among the population indicates that it has long been a human highway, with large numbers of migrants from both directions conquering, trading and generally reproducing along its entire length. The same team also found out that the Bedouin nomads of the Middle East actually have some European and South Asian blood.

In ancient times a Celtic people known as Galatians settled in what is present day Turkey and thus interbred with the people there. During the empire of Alexander the Great, many Greek/Macedonian soldiers had inter-ethnic relationships with women throughout the Middle East all the way to northwestern India. Later, coastal North Africa and parts of the Middle East were part of the Roman Empire and many Roman men were posted as soldiers there. Many of them had inter-ethnic relationships with local Middle-Eastern women. The Germanic people known as the Vandals also conquered parts of coastal North Africa during the great migration period which led to opportunities of inter-ethnic relationships between Germanic men and local North African women.

Inter-ethnic sexual slavery was common during the Arab slave trade throughout the Middle Ages and early modern period, when women and girls captured from non-Arab lands often ended up as sexual slaves in the harems of the Arab World. Most of these slaves came from places such as Sub-Saharan Africa (mainly Zanj), South Asia (Hindus), the North Caucasus (mainly Circassians), Central Asia (mainly Tartars), and Central and Eastern Europe (mainly Saqaliba). The Barbary pirates also captured 1,250,000 slaves from Western Europe and North America between the 16th and 19th centuries. It was also common for Arab conquerors, traders and explorers to marry local females in the lands they conquered or traded with, in various different parts of Africa, Asia.

Inter-ethnic relationships were generally accepted in Arabic society and formed a fairly common theme in medieval Arabic literature and Persian literature. For example, the Kurdish poet Nizami, who had himself married his Kipchak slave girl, wrote The Seven Beauties (1196). Its frame story involves a Persian prince marrying seven foreign princesses, including Byzantine, Chinese, Indian, Khwarezmian, Maghrebian, Slavic and Tartar princesses. Hadith Bayad wa Riyad, a 12th-century Arabic tale from Al-Andalus, was a love story involving an Iberian girl and a Damascene man. The One Thousand and One Nights tale of "The Man of Al-Yaman and His Six Slave-Girls" involves a Yemeni man's relationship with foreign slave girls, four of which are white, black, brown and yellow. Another One Thousand and One Nights tale, "The Ebony Horse", involves the Prince of Persia, Qamar al-Aqmar, rescuing his lover, the Princess of Sana'a, from the Byzantine Emperor who also wishes to marry her.

One study found that some Arabic-speaking populations—Palestinians, Jordanians, Syrians, Iraqis, and Bedouins—have what appears to be substantial mtDNA gene flow from sub-Saharan Africa, amounting to 10–15% of lineages within the past three millennia. In the case of Yemenites, the average is higher at 35%. Of particular historic interest might be the finding that with almost no exceptions the sub-Saharan gene flow was exclusively female, at least in part as a result of the Arab slave trade.

Between the 11th and 13th centuries, medieval Western Asia was repeatedly invaded by Europeans (Crusades) and Mongols (Mongol Empire) which led to opportunities for inter-ethnic relationships between European, Mongol and Central Asian soldiers with local Arab women. As well as many Europeans, there were North Africans, South Asians and Central Asians who worked as mercenaries and traders in the area, most of them converting to Islam and taking local women as wives. From 839 AD, Viking Varangian mercenaries who were in the service of the Byzantine Empire, notably Harald Sigurdsson, campaigned in North Africa, Jerusalem and other places in the Middle East during the Byzantine-Arab Wars, and interbred with the local population as spoils of warfare or through eventual settling with many Scandinavian and Slavic Viking men taking Arab or Anatolian women as wives. There is archaeological evidence these Vikings had established contact with the city of Baghdad, at the time the center of the Islamic Empire, and connected with the populace there.

A genetic anthropological study known as The Genographic Project has found what is believed to be faint genetic traces left by medieval Crusaders in the Middle East. The team has uncovered a specific DNA signature in Lebanon that is probably linked to the Christian crusades of the 7th and 8th centuries. It is believed the Crusaders were welcomed by Christian Arabs long suffering under Islamic rule and many offered their daughters in marriage to the European Crusaders who originated from European kingdoms mostly France, England and the Holy Roman Empire.

In the Ottoman Empire, in addition to the Ottoman elites often taking large numbers of European wives and concubines (see Southeastern and Eastern Europe section), there were also opportunities for the reverse, when the empire recruited young Christian boys (mostly Europeans in addition to Christian Arabs) to become the elite troop of the Turkish Empire, the Janissaries. These Janissaries were stationed throughout the Turkish empire including the Middle-East and North Africa leading to inter-ethnic relationships between European men and women from the Middle East and North Africa. Further inter-ethnic marriages took place during the European colonial age when Great Britain, France and Italy ruled various parts of this region. The majority of these marriages were between European men and local Arab and Berber women.

The concubines of the Ottoman Sultan consisted chiefly of purchased slaves. Because Islamic law forbade Muslims to enslave fellow Muslims, the Sultan's concubines were generally of Christian origin. The mother of a Sultan, though technically a slave, received the extremely powerful title of Valide Sultan, and at times became effective ruler of the Empire. One notable example was Kösem Sultan, daughter of a Greek Christian priest, who dominated the Ottoman Empire during the early decades of the 17th century. Another notable example was Roxelana, the favourite wife of Suleiman the Magnificent.

Inter-ethnic sexual slavery still continues today in a smaller form in the Arab states of the Persian Gulf, where women and children are trafficked from the post-Soviet states, Eastern Europe, Far East, Africa, South Asia and other parts of the Middle East.

Israel

The modern State of Israel was established as a nation-state for the Jewish people. The Jewish identity contains elements of religion (Judaism), ethnicity, and a sense of a common lineage.

In this sense, Jewish miscegenation could be viewed on two levels; one based on belonging to the Jewish ethnic group or Jewish people, and the other based on the race of a given Jew. Jewish miscegenation based on Jewishness (belonging to the Jewish ethnic group or Jewish people) would be defined on whether one parent is not Jewish, independent of whether either the Jewish or non-Jewish parent are of the same or different races. Racial miscegenation would be defined as the union between a Jew of a given race with a person of a different race, be the other person a Jew or not. Two Jewish people may still be considered "interracial" if those two Jews are of different races, although it would not be considered exogamous in the context of Jewish ethnicity, as both are still Jews.

In Israel, all marriages must be approved by religious celebrants, while civil marriages are legally recognized if performed abroad. Rules governing marriage are based on strict religious guidelines of each religion. By Israeli law, authority over all issues related to Judaism in Israel, including marriage, falls under the Orthodox Chief Rabbinate of Israel. Orthodox Judaism is the only form of Judaism recognized by the state, and marriages performed in Israel by non-Orthodox Rabbis are not recognized.

The Rabbinate prohibits marriage in Israel of halakhic Jews (i.e. people born to a Jewish mother or Jewish by conversion), whether they are Orthodox Jews or not, to partners who are non-Jewish or who are of Jewish descent that runs through the paternal line (i.e. not Jewish according to halakha). As a result, in the state of Israel, people of differing religious traditions cannot legally marry someone in another religion and multi-faith couples must leave the country to get married, most often to Central and Northern Greece·

The only other option in Israel for the marriage of a halakhic Jew (Orthodox or not) to a non-Jew, or for that matter, a Christian to a non-Christian or Muslim to a non-Muslim, is for one partner to formally convert to the other's religion, be it to Judaism (Orthodox only), a Christian denomination (such as Eastern Orthodox or Maronite) or a denomination of Islam (such as Sunni or Shia). As for persons with patrilineal Jewish descent (i.e. not recognized as Jewish according to halakha) who wish to marry a halakhic Jew (i.e. born to a Jewish mother or is Jewish by Orthodox conversion) who is Orthodox or otherwise, is also required to formally convert to Judaism (Orthodox only) or they cannot legally marry. According to a Haaretz article "Justice Ministry drafts civil marriage law for 'refuseniks'" 300,000 people, or 150,000 couples, are affected by marriage restrictions based on the partners' disparate religious traditions or non-halakhic Jewish status.

Israeli law concerns itself with miscegenation based on Jewish ethnicity, not miscegenation based on race. Therefore, there are no restrictions on interracial marriages between Jews of different Jewish ethnic divisions, or between other co-religionists of different races, although social stigma may still exist.

Many Israeli Jews oppose mixed relationships between Jewish women and Arab men. A 2007 opinion survey found that more than half of Israeli Jews believed intermarriage is equivalent to "national treason". A group of 35 Jewish men, known as "Fire for Judaism", in Pisgat Ze'ev have started patrolling the town in an effort to stop Jewish women from dating Arab men.

The municipality of Petah Tikva has also announced an initiative to prevent interracial relationships, providing a telephone hotline for friends and family to "inform" on Jewish girls who date Arab men as well as psychologists to provide counselling. The town of Kiryat Gat launched a school programme in schools to warn Jewish girls against dating local Bedouin men.

In February 2010 Maariv has reported that the Tel Aviv municipality has instituted an official, government-sponsored "counselling program" to discourage Jewish girls from dating and marrying Arab boys. The Times has also reported on a vigilante parents' group policing the Jerusalem neighborhood of Pisgat Ze'ev to intimidate and discourage local Arab-Jewish couples.

The Jewish anti-missionary group Yad L'achim has also performed paramilitary "rescue operations" of Jewish women from non-Jewish husbands and celebrates the "rescued women" on their website. However, according to Halakha, descendants from a Jewish mother are always defined to be Jewish.

Australia

Analysis of the 2006 census reveals that 52% of Aboriginal men and 55% of Aboriginal women were married to non-Aboriginal Australians.

New Zealand

Most people who are entitled to call themselves Maori under the Maori Affairs Amendment Act 1974 are less than half Maori. Two-thirds of Maori babies, half of Pacific babies, and a third of white and Asian babies belonged to more than one ethnic group.

Portuguese Colonies

According to Gilberto Freyre, a Brazilian sociologist, miscegenation was commonplace in the Portuguese colonies, and was even supported by the court as a way to boost low populations and guarantee a successful and cohesive settlement. Thus, settlers often released African slaves to become their wives. The children were guaranteed full Portuguese citizenship, provided the parents were married. Some former Portuguese colonies have large mixed-race populations, for instance, Brazil, Cape Verde, Mozambique, Timor Leste, Macau and São Tomé and Príncipe. In the case of Brazil, the influential "Indianist" novels of José de Alencar (O Guarany, Iracema, and Ubirajara) perhaps went farther than in the other colonies, advocating miscegenation in order to create a truly Brazilian race. Mixed marriages between Portuguese and locals in former colonies were very common in all Portuguese colonies. Miscegenation was still common in Africa until the independence of the former Portuguese colonies in the mid-1970s.

Demographics of Ethnoracial Admixture

U.S.

According to the U.S. Census, in 2000 there were 504,119 Asian-white marriages, 287,576 black-white marriages, and 31,271 Asian-black marriages. The black-white marriages increased from 65,000 in 1970 to 558,000 in 2010, according to Census Bureau figures.

In the United States, rates of interracial cohabitation are significantly higher than those of marriage. Although only 7% of married African American men have Caucasian American wives, 13% of cohabitating African American men have Caucasian American partners. 25% of married Asian American women have Caucasian spouses, but 45% of cohabitating Asian American women are with Caucasian American men. Of cohabiting Asian men, slightly over 37 percent of Asian men have white female partners. In 2006, 41% of Asian American-born women were registered as having White husbands. Almost 30% of Asians and Latinos outmarry, with 86.8 and 90% of these intermarriages, respectively, being to a white person. According to Karyn Langhorne Folan, "...although the most recent census available reported that 70% of African American women are single, African American women have the greatest resistance to marrying 'out' of the race."

One study on college students noted that, while 90% of the 80 black men reported experiences in interracial dating, only 12% of the 140 white men reported such experiences. Some studies have shown as many as 45% of white women have dated men outside their race. The massive sexual survey revealed that 19% of black males had engaged in sexual activity with white women.

A Gallup poll on interracial dating in June 2006 found that 95% of 18- to 29-year-olds approve of blacks' and whites' dating each other. About 60% of that age group said they have dated someone of another race. In 1980, just 17% said they had dated someone from a different racial background.

According to the 2010 Census, the number of interracial marriages in the U.S. is up 20% since 2000, to around 4.5 million. "A record 14.6% of all new marriages in the United States in 2008 were between spouses of a different race or ethnicity from one another.... Rates more than doubled among whites and nearly tripled among blacks. But for both Hispanics and Asians, rates were nearly identical in 2008 and 1980.", according to a Pew Research Center analysis of demographic data from the U.S. Census Bureau.

According to studies by Jenifer L. Bratter and Rosalind B. King made publicly available on the Education Resources Information Center, White female-Black male and White female-Asian male marriages are more prone to divorce than White-White pairings. Conversely, unions between White males and non-White females (and between Hispanics and non-Hispanic persons) have similar or lower risks of divorce than White-White marriages, unions between white-black female last longer than white-white pairings or white-Asian pairings.

Brazil

Multiracial Brazilians make up 42.6% of Brazil's population, 79.782 million people, and they live in all regions of Brazil. Multiracial Brazilians are mainly people of mixed European, African, East Asian (mostly Japanese) and Amerindian ancestry.

Genetic Studies of Racial Admixture

Miscegenation between two populations reduces the genetic distance between the populations. During the Age of Discovery which began in the early 15th century, European explorers sailed all across the globe reaching all the major continents. In the process they came into contact with many populations that had been isolated for thousands of years. The Tasmanian aboriginals were one of the most isolated groups on the planet. They were driven to extinction by European explorers, however a number of their descendants survive today as a result of admixture with Europeans. This is an example of how modern migrations have begun to reduce the genetic divergence of the human race.

The demographic composition of the old world has not changed significantly since the age of discovery. However, the new world demographics were radically changed within a short time following the voyage of Columbus. The colonization of Americas brought Native Americans into contact with the distant populations of Europe, Africa and Asia. As a result many countries in the Americas have significant and complex multiracial populations. Furthermore many who identify themselves by only one race still have multiracial ancestry.

Admixture in the United States

***Table:** Admixture in European-American population*

% European admixture	*Frequency*
90-100	68%
80-89.9	22%
70-79.9	8%
60-69.9	< 1%
50-59.9	< 1%
40-49.9	< 1%
0-39.9	0

Some claim the vast majority of African-Americans possess varying degrees of European admixture (the average Black American is 20% European) although studies suggest the Native American admixture in Black Americans is highly exaggerated; some estimates put average African-American possession of European admixture at 25% with figures as high as 50% in the Northeast and less than 10% in the south.

A recent study by Mark D. Shriver of a European-American sample found that the average admixture in the white population is 0.7% African and 3.2% Native American. However, 70% of the sample had no African admixture. The other 30% had African admixture ranging from 2% to 20% with an average of 2.3%. By extrapolating these figures to the whole population some scholars suggest that up to 74 million European-Americans may have African admixture in the same range (2–20%).

Dr Mark Shriver, the team leader of the study, found that he had 11% West African ancestry though he identifies as white. Studies based on skin reflectance have shown the color line in the US applied selective pressure on genes that code for skin color but did not apply any selective pressure on other invisible African genes. Since there are an estimated 6 genetic loci involved in skin color determination it is possible for someone to have 15–20% African admixture and not possess any of alleles that code for dark skin. This is the basis of the passing phenomenon.

Thus African admixture amongst white Americans can increase without any significant change in skin tone. Conversely amongst African-Americans, an amount of African Admixture is directly correlated with darker skin since no selective pressure is applied; as a result, African-Americans may have a much wider range of African admixture (>0–100%), whereas European-

Americans have a lower range (2–20%). A small overlap exists so that it is possible that someone who identifies himself as white may have more African admixture than a person who identifies himself as black.

A statistical analysis done in 1958 using historical census data and historical data on immigration and birth rates, concluded that 21 percent of the white population had black ancestors. The growth in the white population could not be attributed to births in the white population and immigration from Europe alone, but had received significant contribution from the African American population as well. The author states in 1958:

The data presented in this study indicate that the popular belief in the non-African background of white persons is invalid. Over twenty-eight million white persons are descendants of persons of African origin. Furthermore, the majority of the persons with African ancestry are classified as white.

In the United States intermarriage among Filipinos with other races is common. They have the largest number of interracial marriages among Asian immigrant groups, as documented in California. It is also noted that 21.8% of Filipino Americans are of mixed blood, second among Asian Americans, and is the fastest growing.

Admixture in Latin America

Prior to the European conquest of the Americas the demographics of Latin America was naturally 100% American Indian. Today those who identify themselves as Native Americans are small minorities in many countries. For example the CIA lists Argentina's native population at 0.9%, Brazil's at 0.4%, and Uruguay's at 0%.

The early conquest of Latin America was primarily carried out by male soldiers and sailors from Spain and Portugal. Since they carried very few European women on their journeys the new settlers married and fathered children with Amerindian women and also with women imported from Africa. This process of miscegenation was even encouraged by the Spanish monarchy and it led to the system of stratification known as the Casta.

This system had Europeans (Spaniards and Portuguese) at the top of the hierarchy followed by those of mixed race. Unmixed Blacks and Native Americans were at the bottom. A philosophy of whitening emerged in which Amerindian and African culture was stigmatized in favor of European values.

Many Amerindian languages were lost as mixed race offspring adopted Spanish and Portuguese as their first languages. Only towards the end of the 19th Century and beginning of the 20th century did large numbers of Europeans begin to migrate to South America and consequently altering its demographics. In addition many Africans were shipped to regions all over the Americas and were present in many of the early voyages of the conquistadors. Brazil has the largest population of African descendants outside Africa. Other countries such as Jamaica, Cuba, Puerto Rico, Dominican Republic, Haiti, Venezuela, Colombia, and Ecuador still have sizeable populations identified as Black. However countries such as Argentina and Chile do not have a visible African presence today. Census information from the early 19th century shows that people categorized as Black made up to 30% of the population, or around 400,000 people.

Though almost completely absent today, their contribution to Argentine culture is significant include the tango, the milonga and the zamba, words of Bantu origin.

Table: *Demographics of Brazil in 1835, 1940, 2000 and 2008*

Year	*White*	*Brown*	*Black*
1835	24.4%	18.2%	51.4%
1940	64%	21%	14%
2000	53.7%	38.5%	6.2%
2008	48.8%	43.8%	6.5%

The ideology of whitening encouraged non-whites to seek white or lighter skinned partners. This dilution of non-white admixture would be beneficial to their offspring as they would face less stigmatization and find it easier to assimilate into mainstream society. After successive generations of European gene flow, non-white admixture levels would drop below levels at which skin color or physical appearance is not affected thus allowing individuals to identify as white. In many regions, the native and black populations were simply overwhelmed by a succession of waves of European immigration.

Historians and scientists are thus interested in tracing the fate of Native Americans and Africans from the past to the future. The questions remain about what proportion of these populations simply died out and what proportion still has descendants alive today including those who do not racially identify themselves as their ancestors would have. Admixture testing has thus become a useful objective tool in shedding light on the demographic history of Latin America.

Recent Studies

Unlike in the United States, there were no anti-miscegenation policies in Latin America. Though still a racially stratified society there were no significant barriers to gene flow between the three populations. As a result admixture profiles are a reflection of the colonial populations of Africans, Europeans and Amerindians. The pattern is also sex biased in that the African and Amerindian maternal lines are found in significantly higher proportions than African or

Amerindian Y chromosomal lines. This is an indication that the primary mating pattern was that of European males with Amerindian or African females. According to the study more than half the white populations of the Latin American countries studied have some degree of either native American or African admixture (MtDNA or Y Chromosome). In countries such as Chile and Colombia almost the entire white population was shown to have some non-white admixture Following the dispersal of Humans from Africa 50,000 – 70,000 years ago South America was the last continent to be occupied by humans. Thus the largest geographic distance between continents is between Africa and South America. Since genetic distance increases with geographic distance the two most genetically divergent groups are Africans and Native Americans based on distance. The arrival of Africans in Brazil and subsequent mixing with native South Americans entails the creation of intermediate populations, such as the Zambo or Garifuna between the two divergent groups.

Frank Moya Pons, a Dominican historian documented that Spanish colonists intermarried with Taíno women, and, over time, these mestizo descendants intermarried with Africans, creating a tri-racial Creole culture. 1514 census records reveal that 40% of Spanish men in the colony of Santo Domingo had Taíno wives. A recent study conducted in Puerto Rico suggests that over 61% of the population possess Amerindian mtDNA.

Admixture in the Philippines

Historically, admixture has been an ever present and pervading phenomenon in the Philippines. The Philippines was originally settled by Australoid peoples called Negritos which now form the country's aboriginal community. Admixture occurred between this earlier group and the mainstream Malayo-Polynesian population.

There has been Indian migration to and influence in the Philippines since the precolonial era. About 25% of the words in the Tagalog language are Sanskrit terms and about 5% of the country's population possess Indian ancestry from antiquity. There has been a Chinese presence in the Philippines since the 9th century. However, large-scale migrations of Chinese to the Philippines only started during the Spanish colonial era, when the world market was opened to the Philippines. It is estimated that among Filipinos, 10%–20% have some Chinese ancestry and 1.5% are "full-blooded" Chinese.

According to the American anthropologist Dr. H. Otley Beyer, the ancestry of Filipinos is 2% Arab. This dates back to when Arab traders intermarried with the local Malay Filipina female populations during the pre-Spanish history of the Philippines. A recent genetic study by Stanford University indicates that at least 3.6% of the population are European or of part European descent from both Spanish and United States colonization.

Admixture Among the Romani People

Genetic evidence has shown that the Romani people ("Gypsies") originated from the Indian subcontinent and mixed with the local populations in Central Asia, the Middle East, and Europe. In the 1990s, it was discovered that Romani populations carried large frequencies of particular Y chromosomes (inherited paternally) that otherwise exist only in populations from South Asia, in addition to fairly significant frequencies of particular mitochondrial DNA (inherited maternally) that is rare outside South Asia.

47.3% of Romani males carry Y chromosomes of haplogroup H-M82 which is rare outside of the Indian subcontinent. Mitochondrial haplogroup M, most common in Indian subjects and rare outside Southern Asia, accounts for nearly 30% of Romani people. A more detailed study of Polish Roma shows this to be of the M5 lineage, which is specific to India. Moreover, a form of the inherited disorder congenital myasthenia is found in Romani subjects. This form of the disorder, caused by the 1267delG mutation, is otherwise only known in subjects of Indian ancestry. This is considered to be the best evidence of the Indian ancestry of the Romanies.

The Romanis have been described as "a conglomerate of genetically isolated founder populations", while a number of common Mendelian disorders among Romanies from all over Europe indicates "a common origin and founder effect".

A study from 2001 by Gresham et al. suggests "a limited number of related founders, compatible with a small group of migrants splitting from a distinct caste or tribal group". Also the study pointed out that "genetic drift and different levels and sources of admixture, appear to have played a role in the subsequent differentiation of populations". The same study found that "a single lineage... found across Romani populations, accounts for almost one-third of Romani males. A similar preservation of a highly resolved male lineage has been reported elsewhere only for Jewish priests".

A 2004 study by Morar et al. concluded that the Romani are "a founder population of common origins that has subsequently split into multiple socially divergent and geographically dispersed Gypsy groups". The same study revealed that this population "was founded approximately 32–40 generations ago, with secondary and tertiary founder events occurring approximately 16–25 generations ago".

Hybrid

In biology and specifically, genetics, the term hybrid has several meanings, all referring to the offspring of sexual reproduction.

1. In general usage, hybrid is synonymous with heterozygous: any offspring resulting from the mating of two distinctly homozygous individuals

2. a genetic hybrid carries two different alleles of the same gene
3. a structural hybrid results from the fusion of gametes that have differing structure in at least one chromosome, as a result of structural abnormalities
4. a numerical hybrid results from the fusion of gametes having different haploid numbers of chromosomes
5. a permanent hybrid is a situation where only the heterozygous genotype occurs, because all homozygous combinations are lethal.

From a taxonomic perspective, hybrid refers to offspring resulting from the interbreeding between two animals or plants of different taxa.

1. Hybrids between different subspecies within a species (such as between the Bengal tiger and Siberian tiger) are known as intra-specific hybrids. Hybrids between different species within the same genus (such as between lions and tigers) are sometimes known as interspecific hybrids or crosses. Hybrids between different genera (such as between sheep and goats) are known as intergeneric hybrids. Extremely rare interfamilial hybrids have been known to occur (such as the guineafowl hybrids). No interordinal (between different orders) animal hybrids are known.
2. The second type of hybrid consists of crosses between populations, breeds or cultivars within a single species. This meaning is often used in plant and animal breeding, where hybrids are commonly produced and selected because they have desirable characteristics not found or inconsistently present in the parent individuals or populations. This flow of genetic material between populations or races is often called hybridization.

Etymology

According to the Oxford English Dictionary, the word is derived from Latin hybrida , meaning the "offspring of a tame sow and a wild boar", "child of a freeman and slave", etc. The term entered into popular use in English in the 19th century, though examples of its use have been found from the early 17th century.

Types of Hybrids

Depending on the parents, there are a number of different types of hybrids;

- Single cross hybrids — result from the cross between two true breeding organisms and produces an F1 generation called an F1 hybrid (F1 is short for Filial 1, meaning "first offspring"). The cross between two different homozygous lines produces an F1 hybrid that is heterozygous;

having two alleles, one contributed by each parent and typically one is dominant and the other recessive. The F1 generation is also phenotypically homogeneous, producing offspring that are all similar to each other.

- Double cross hybrids — result from the cross between two different F1 hybrids.
- Three-way cross hybrids — result from the cross between one parent that is an F1 hybrid and the other is from an inbred line.
- Triple cross hybrids — result from the crossing of two different three-way cross hybrids.
- Population hybrids — result from the crossing of plants or animals in a population with another population. These include crosses between organisms such as interspecific hybrids or crosses between different races.

Interspecific Hybrids

Interspecific hybrids are bred by mating two species, normally from within the same genus. The offspring display traits and characteristics of both parents. The offspring of an interspecific cross are very often sterile; thus, hybrid sterility prevents the movement of genes from one species to the other, keeping both species distinct. Sterility is often attributed to the different number of chromosomes the two species have, for example donkeys have 62 chromosomes, while horses have 64 chromosomes, and mules and hinnies have 63 chromosomes. Mules, hinnies, and other normally sterile interspecific hybrids cannot produce viable gametes, because, the extra chromosome cannot make a homologous pair at meiosis, meiosis is disrupted, and viable sperm and eggs are not formed. However, fertility in female mules has been reported with a donkey as the father.

Most often other processes occurring in plants and animals keep gametic isolation and species distinction. Species often have different mating or courtship patterns or behaviors, the breeding seasons may be distinct and even if mating does occur antigenic reactions to the sperm of other species prevent fertilization or embryo development.

The Lonicera fly is the first known animal species that resulted from natural hybridization. Until the discovery of the Lonicera fly, this process was known to occur in nature only among plants. While it is possible to predict the genetic composition of a backcross on average, it is not possible to accurately predict the composition of a particular backcrossed individual, due to random segregation of chromosomes. In a species with two pairs of chromosomes, a twice backcrossed individual would be predicted to contain 12.5% of one species' genome (say, species A). However, it may, in fact, still be a 50% hybrid if the chromosomes from species A were lucky in two successive segregations, and meiotic crossovers happened near the telomeres. The chance of this is fairly high: $\left(\frac{1}{2}\right)^{(2\times2)} = \frac{1}{16}$ (where the "two times two" comes about from two rounds of meiosis with two chromosomes); however, this probability declines markedly with chromosome number and so the actual composition of a hybrid will be increasingly closer to the predicted composition.

Hybrids are often named by the portmanteau method, combining the names of the two parent species. For example, a zeedonk is a cross between a zebra and a donkey. Since the traits of hybrid offspring often vary depending on which species was mother and which was father, it is traditional to use the father's species as the first half of the portmanteau. For example, a liger is a cross between a male lion and a female tiger, while a tiglon is a cross between a male tiger and a female lion.

Domestic and Wild Hybrids

This section does not cite any references or sources. Please help improve this section by adding citations to reliable sources. Unsourced material may be challenged and removed. (January 2010)

Hybrids between domesticated and wild animals in particular may be problematic. Breeders of domesticated species discourage crossbreeding with wild species, unless a deliberate decision is made to incorporate a trait of a wild ancestor back into a given breed or strain. Wild populations of animals and plants have evolved naturally over millions of years through a process of natural selection in contrast to human controlled selective breeding or artificial selection for desirable traits from the human point of view. Normally, these two methods of reproduction operate independently of one another.

However, an intermediate form of selective breeding, wherein animals or plants are bred by humans, but with an eye to adaptation to natural region-specific conditions and an acceptance of natural selection to weed out undesirable traits, created many ancient domesticated breeds or types now known as landraces. Many times, domesticated species live in or near areas which also still hold naturally evolved, region-specific wild ancestor species and subspecies. In some cases, a domesticated species of plant or animal may become feral, living wild. Other times, a wild species will come into an area inhabited by a domesticated species. Some of these situations lead to the creation of hybridized plants or animals, a cross between the native species and a domesticated one. This type of crossbreeding, termed genetic pollution by those who are concerned about preserving the genetic base of the wild species, has become a major concern. Hybridization is also a concern to the breeders of purebred species as well, particularly if the gene pool is small and if such crossbreeding or hybridization threatens the genetic base of the domesticated purebred population.

The concern with genetic pollution of a wild population is that hybridized animals and plants may not be as genetically strong as naturally evolved region specific wild ancestors wildlife which can survive without human husbandry and have high immunity to natural diseases. The concern of purebred breeders with wildlife hybridizing a domesticated species is that it can coarsen or degrade the specific qualities of a breed developed for a specific purpose, sometimes over many generations. Thus, both purebred breeders and wildlife biologists share a common interest in preventing accidental hybridization.

Examples of Hybrid Animals

Figure: *A "Zeedonk", a zebra/donkey hybrid*

Figure: *Hercules, a "Liger", a Lion/Tiger hybrid*

Figure: *A "Jaglion", a Jaguar/Lion hybrid*

Figure: *A mule, a Domestic Canary/Goldfinch hybrid.*

- Hybrid Iguana, a single cross hybrid resulting from natural interbreeding between male marine iguanas and female land iguanas since the late 2000s.
- Equid hybrids
  - o Mule, a cross of female horse and a male donkey.
  - o Hinny, a cross between a female donkey and a male horse. Mule and hinny are examples of reciprocal hybrids.
  - o Zebroids
    - § Zeedonk or Zonkey, a zebra/donkey cross.
    - § Zorse, a zebra/horse cross
    - § Zony or Zetland, a zebra/pony cross ("zony" is a generic term; "zetland" is specifically a hybrid of the Shetland pony breed with a zebra)
- Bovid hybrids
  - o Dzo, zo or yakow; a cross between a domestic cow/bull and a yak.
  - o Beefalo, a cross of an American bison and a domestic cow. This is a fertile breed; this along with genetic evidence has caused them to be recently reclassified into the same genus, Bos.
  - o Zubron, a hybrid between wisent (European bison) and domestic cow.
- Sheep-goat hybrids, such as the Toast of Botswana.
- Ursid hybrids, such as the grizzly-polar bear hybrid, occur between black bears, brown bears, and polar bears.
- Felid hybrids
  - o Savannah cats are the hybrid cross between an African serval cat and a domestic cat
  - o A hybrid between a Bengal tiger and a Siberian tiger is an example of an intra-specific hybrid.

- o Ligers and Tiglons (crosses between a lion and a tiger) and other Panthera hybrids such as the lijagulep. Various other wild cat crosses are known involving the lynx, bobcat, leopard, serval, etc.
- o Bengal cat, a cross between the Asian leopard cat and the domestic cat, one of many hybrids between the domestic cat and wild cat species. The domestic cat, African wild cat and European wildcat may be considered variant populations of the same species (Felis silvestris), making such crosses non-hybrids.

- Fertile canid hybrids occur between coyotes, wolves, dingoes, jackals and domestic dogs.
- Hybrids between black and white rhinoceroses have been recognized.
- Hybrids between spotted owls and barred owls
- Cama, a cross between a camel and a llama, also an intergeneric hybrid.
- Wholphin, a fertile but very rare cross between a false killer whale and a bottlenose dolphin.
- A fertile cross between a king snake and a corn snake.
- At Chester Zoo in the United Kingdom, a cross between an African elephant (male) and an Asian elephant (female). The male calf was named Harjeeven. It died of intestinal infection after twelve days.
- Cagebird breeders sometimes breed hybrids between species of finch, such as goldfinch x canary. These birds are known as mules.
- Gamebird hybrids, hybrids between gamebirds and domestic fowl, including chickens, guineafowl and peafowl, interfamilial hybrids.
- Numerous macaw hybrids are also known.
- Red kite x black kite: five bred unintentionally at a falconry center in England. (It is reported that the black kite (the male) refused female black kites but mated with two female red kites.)
- Hybridization between the endemic Cuban crocodile (Crocodilus rhombifer) and the widely distributed American crocodile (Crocodilus acutus) is causing conservation problems for the former species as a threat to its genetic integrity.
- Saltwater crocodiles (Crocodylus porosus) have mated with Siamese crocodiles (Crocodylus siamensis) in captivity producing offspring which in many cases have grown over 20 feet (6 meters) in length. It is likely that wild hybridization occurred historically in parts of southeast Asia.
- Blood parrot cichlid, which is probably created by crossing a red head cihclid and a Midas cichlid or red devil cichlid

- The mulard duck, hybrid of the domestic pekin duck and domesticated muscovy ducks.
- Killer bees were created in an attempt to breed tamer and more manageable bees. This was done by crossing a european honey bee and an african bee, but instead the offspring became more aggressive and highly defensive bees that had escaped into the wild.
- In Australia, New Zealand and other areas where the Pacific Black Duck occurs, it is hybridised by the much more aggressive introduced Mallard. This is a concern to wildlife authorities throughout the affected area, as it is seen as Genetic pollution of the Black Duck gene pool.

Hybrids should not be confused with genetic chimeras such as that between sheep and goat known as the geep. Wider interspecific hybrids can be made via in vitro fertilization or somatic hybridization, however the resulting cells are not able to develop into a full organism. An example of interspecific hybrid cell lines is humster (hamster x human) cells.

Hybrid Plants

Plant species hybridize more readily than animal species, and the resulting hybrids are more often fertile hybrids and may reproduce, though there still exist sterile hybrids and selective hybrid elimination where the offspring are less able to survive and are thus eliminated before they can reproduce.

A number of plant species are the result of hybridization and polyploidy with many plant species easily cross pollinating and producing viable seeds, the distinction between each species is often maintained by geographical isolation or differences in the flowering period. Since plants hybridize frequently without much work, they are often created by humans in order to produce improved plants.

Figure: *A sterile Trillium hybrid between Trillium cernuum and Trillium grandiflorum.*

These improvements can include the production of more or improved; seeds, fruits or other plant parts for consumption, or to make a plant more winter or heat hardy or improve its growth and/or appearance for use in horticulture. Much work is now being done with hybrids to produce more disease resistant plants for both agricultural and horticultural crops. In many groups of plants hybridization has been used to produce larger and more showy flowers and new flower colors.

Many plant genera and species have their origins in polyploidy. Autopolyploidy resulting from the sudden multiplication in the number of chromosomes in typical normal populations caused by unsuccessful separation of the chromosomes during meiosis. Tetraploids or plants with four sets of chromosomes are common in a number of different groups of plants and over time these plants can differentiate into distinct species from the normal diploid line. In Oenothera lamarchiana the diploid species has 14 chromosomes, this species has spontaneously given rise to plants with 28 chromosomes that have been given the name Oenthera gigas. Tetraploids can develop into a breeding population within the diploid population and when hybrids are formed with the diploid population the resulting offspring tend to be sterile triploids, thus effectively stopping the intermixing of genes between the two groups of plants (unless the diploids, in rare cases, produce unreduced gametes).

Figure: *An ornamental lily hybrid known as Lilium 'Citronella'*

Another form of polyploidy called allopolyploidy occurs when two different species mate and produce hybrids. Usually the typical chromosome number is doubled in successful allopolyploid species, with four sets of chromosomes the genotypes can sort out to form a complete diploid set from the parent species, thus they can produce fertile offspring that can mate and reproduce with each other but can not back-cross with the parent species. Allopolyploidy in plants often gives them a condition called hybrid vigour, which results in plants that are larger and stronger growing than either of the two parent species. Allopolyploids are often more aggressive growing and can be invaders of new habitats. Sterility in a hybrid is often a result of chromosome number;

if parents are of differing chromosome pair number, the offspring will have an odd number of chromosomes, leaving them unable to produce chromosomally balanced gametes. While this is a negative in a crop such as wheat, when growing a crop which produces no seeds would be pointless, it is an attractive attribute in some fruits. Bananas and seedless watermelon, for instance, are intentionally bred to be triploid, so that they will produce no seeds. Many hybrids are created by humans, but natural hybrids occur as well.

Heterosis

Hybrids are sometimes stronger than either parent variety, a phenomenon most common with plant hybrids, which when present is known as hybrid vigor (heterosis) or heterozygote advantage. A transgressive phenotype is a phenotype displaying more extreme characteristics than either of the parent lines. Plant breeders make use of a number of techniques to produce hybrids, including line breeding and the formation of complex hybrids. An economically important example is hybrid maize (corn), which provides a considerable seed yield advantage over open pollinated varieties. Hybrid seed dominates the commercial maize seed market in the United States, Canada and many other major maize producing countries.

Examples of Species Hybrids

Some plant hybrids include:

- Leyland cypress, [x Cupressocyparis leylandii] hybrid between Monterey Cypress and Nootka Cypress.
- Limequat, lime and kumquat hybrid.
- Loganberry, a hybrid between raspberry and blackberry.
- London plane, a hybrid between Platanus orientalis Oriental plane and Platanus occidentalis American plane (American sycamore). Thus forming [Platanus x acerifolia]
- Magnolia × alba, an interspecific hybrid between Magnolia champaca and Magnolia montana
- Peppermint, a hybrid between spearmint and water mint.
- Tangelo, a hybrid of a Mandarin orange and a pomelo which may have been developed in Asia about 3,500 years ago.
- Triticale, a wheat–rye hybrid.
- Wheat; most modern and ancient wheat breeds are themselves hybrids. Bread wheat is a hexaploid hybrid of three wild grasses; durum (pasta) wheat is a tetraploid hybrid of two wild grasses.
- Triangle of U: cabbage, mustard etc.
- Grapefruit, hybrid between a pomelo and the Jamaican sweet orange.

Some natural hybrids are:

- Iris albicans, a sterile hybrid which spreads by rhizome division
- Evening primrose, a flower which was the subject of famous experiments by Hugo de Vries on polyploidy and diploidy.

Some horticultural hybrids:

- Dianthus x allwoodii, is a hybrid between Dianthus caryophyllus × Dianthus plumarius. This is an "interspecific hybrid" or hybrid between two species in the same genus.
- x Heucherella tiarelloides, or Heuchera sanguinea × Tiarella cordifolia is an "intergeneric hybrid" a hybrid between two different genera.
- Quercus x warei [Quercus robur x Quercus bicolor] Kindred Spirit Hybrid Oak

Hybrids in Nature

Hybridisation between two closely related species is actually a common occurrence in nature but is also being greatly influenced by anthropogenic changes as well. Hybridization is a naturally occurring genetic process where individuals from two genetically distinct populations mate. As stated above, it can occur both intraspecifically, between different distinct populations within the same species, and interspecifically, between two different species. Hybrids can be either sterile/not viable or viable/fertile.

This affects the kind of effect that this hybrid will have on its and other populations that it interacts with. Many hybrid zones are known where the ranges of two species meet, and hybrids are continually produced in great numbers. These hybrid zones are useful as biological model systems for studying the mechanisms of speciation (Hybrid speciation). Recently DNA analysis of a bear shot by a hunter in the North West Territories confirmed the existence of naturally-occurring and fertile grizzly–polar bear hybrids. There have been reports of similar supposed hybrids, but this is the first to be confirmed by DNA analysis. In 1943, Clara Helgason described a male bear shot by hunters during her childhood. It was large and off-white with hair all over its paws. The presence of hair on the bottom of the feet suggests it was a natural hybrid of Kodiak and Polar bear.

Anthropogenic Hybridization

Anthropogenic caused changes to the environment such as fragmentation and Introduced species are becoming more widespread. This is allowing another kind of hybridization that is more of the focus of conservation genetics to occur: anthropogenic hybridization. This anthropogenically caused hybridization increases the challenges in managing certain populations that are experiencing introgression.

5

Clinical Trials

On January 23, 2009, the US Food and Drug Administration gave clearance to Geron Corporation for the initiation of the first clinical trial of an embryonic stem cell-based therapy on humans. The trial will evaluate the drug GRNOPC1, embryonic stem cell-derived oligodendrocyte progenitor cells, on patients with acute spinal cord injury. As of mid 2010 hundreds of phase III clinical trials involving stem cells have been registered.

Stem Cell Use in Animals

Veterinary Applications Potential Contributions to Veterinary Medicine

Research currently conducted on horses, dogs, and cats can benefit the development of stem-cell treatments in veterinary medicine and can target a wide range of injuries and diseases such as myocardial infarction, stroke, tendon and ligament damage, osteoarthritis, osteochondrosis and muscular dystrophy both in large animals, as well as humans. While investigation of cell-based therapeutics generally reflects human medical needs, the high degree of frequency and severity of certain injuries in racehorses has put veterinary medicine at the forefront of this novel regenerative approach. Companion animals can serve as clinically relevant models that closely mimic human disease.

Development of Regenerative Treatment Models

Veterinary applications of stem cell therapy as a means of tissue regeneration have been largely shaped by research that began with the use of adult-derived mesenchymal stem cells to treat animals with injuries or defects affecting bone, cartilage, ligaments and/or tendons. Because mesenchymal stem cells can differentiate into the cells that make up bone, cartilage, tendons, and ligaments (as well as muscle, fat, and possibly other tissues), they have been the main type of stem cells studied in the treatment of diseases affecting these tissues. Mesenchymal stem cells are primarily derived from adipose tissue or bone marrow. Since an elevated immune response following cell transplantation may result in rejection of exogenous cells (except in the case of cells derived from a very closely genetically related individual), mesenchymal stem cells are often derived from the patient prior to injection in a process known as autologous transplantation. Surgical repair of bone fractures in dogs and sheep has demonstrated that engraftment of mesenchymal stem cells derived from a genetically different donor within the same species, termed allogeneic transplantation, does not elicit an immunological response in the recipient animal and can mediate regeneration of bone tissue in major bony fractures and defects.

Stem cells can speed up bone repair in fractures/defects that would normally require extensive grafting, suggesting that mesenchymal stem cell use may provide a useful alternative to conventional grafting techniques. Treating tendon and ligament injuries in horses using stem cells, whether derived from adipose tissue or bone-marrow, has support in the veterinary literature. While further studies are necessary to fully characterize the use of cell-based therapeutics for treatment of bone fractures, stem cells are thought to mediate repair via five primary mechanisms: 1) providing an antiinflammatory effect, 2) homing to damaged tissues and recruiting other cells, such as endothelial progenitor cells, that are necessary for tissue growth, 3) supporting tissue remodeling over scar formation, 4) inhibiting apoptosis, and 5) differentiating into bone, cartilage, tendon, and ligament tissue.

Significance of Stem Cell Microenvironments

The microenvironment into which stem cells are transplanted significantly alters the capacity of engrafted cells for recovery and repair. The microenviroment provides growth factors and other chemical signals that guide appropriate differentiation of transplanted cell populations and direct transplanted cells to sites of trauma or disease. Repair and recovery can then be mediated via three primary mechanisms: 1) formation and/or recruitment of new blood cells to the damaged region; 2) prevention of programed cell death or apoptosis; and 3) suppression of inflammation. To further enrich blood supply to the damaged areas, and consequently promote tissue regeneration, platelet-rich plasma could be used in conjunction with stem cell transplantation. The efficacy of some stem cell populations may also be affected by the method of delivery; for instance, to regenerate bone, stem cells are often introduced in a scaffold where they produce the minerals necessary for generation of functional bone.

Sources of Autologous (Patient-derived) Stem Cells

Autologous stem cells intended for regenerative therapy are generally isolated either from the patient's bone marrow or from adipose tissue. The number of stem cells transplanted into damaged tissue may alter efficacy of treatment.

Accordingly, stem cells derived from bone marrow aspirates, for instance, are cultured in specialized laboratories for expansion to millions of cells. Although adipose-derived tissue also requires processing prior to use, the culturing methodology for adipose-derived stem cells is not as extensive as that for bone marrow-derived cells. While it is thought that bone-marrow derived stem cells are preferred for bone, cartilage, ligament, and tendon repair, others believe that the less challenging collection techniques and the multi-cellular microenvironment already present in adipose-derived stem cell fractions make the latter the preferred source for autologous transplantation.

Currently available treatments for horses and dogs suffering from orthopaedic conditions Autologous or allogeneic stem cells are currently used as an adjunctive therapy in the surgical repair of some types of fractures in dogs and horses. Autologous stem cell-based treatments for ligament injury, tendon injury, osteoarthritis, osteochondrosis, and sub-chondral bone cysts have been commercially available to practicing veterinarians to treat horses since 2003 in the United States and since 2006 in the United Kingdom. Autologous stem-cell based treatments for tendon injury, ligament injury, and osteoarthritis in dogs have been available to veterinarians in the United States since 2005. Over 3000 privately-owned horses and dogs have been treated with autologous adipose-derived stem cells. The efficacy of these treatments has been shown in double-blind clinical trials for dogs with osteoarthritis of the hip and elbow and horses with tendon damage. The efficacy of using stem cells, whether adipose-derived or bone-marrow derived, for treating tendon and ligament injuries in horses has support in the veterinary literature.

Developments in Stem Cell Treatments in Veterinary Internal Medicine Currently, research is being conducted to develop stem cell treatments for: 1) horses suffering from COPD, neurologic disease, and laminitis; and 2) dogs and cats suffering from heart disease, liver disease, kidney disease, neurologic disease, and immune-mediated disorders.

Embryonic Stem Cell Controversy

There is widespread controversy over the use of human embryonic stem cells. This controversy primarily targets the techniques used to derive new embryonic stem cell lines, which often requires the destruction of the blastocyst. Opposition to the use of human embryonic stem cells in research is often based on philosophical, moral or religious objections.

Stem Cell Treatments Around the World

China

Stem cell research and treatment is currently being practiced at a clinical level in the People's Republic of China. The Ministry of Health of the People's Republic of China has permitted the use of stem cell therapy for conditions beyond those approved of in Western countries such as the United States, United Kingdom, and Australia. The Western World has scrutinized China for its failed attempts to meet international documentation standards of these trials and procedures, despite the overwhelmingly positive anecdotal results.

Stem cell therapies provided in China utilize a variety of cell types including umbilical cord stem cells and olfactory ensheathing cells. The stem cells are then expanded in centralized blood banks before being used in stem cell treatments. State-funded companies based in the Shenzhen Hi-Tech Industrial Zone treat the symptoms of numerous disorders with adult stem cell therapy. Hospitals throughout eastern China provide numerous therapies to patients in coordination with the stem cell providers.

These companies' therapies are currently focused on the treatment of neurodegenerative and cardiovascular disorders. The most radical successes of Chinese adult stem cell therapy have been in treating the brain. These therapies administer stem cells directly to the brain to promote greater motor and brain function in patients with Cerebral Palsy, Alzheimer's, and brain injuries.

However, retrospective studies have shown that Chinese use of fetal-derived brain tissue in spinal cord injured human subjects were not as promising as once thought: the phenotype and the fate of the transplanted cells, described as olfactory ensheathing cells, were unknown. As well, perioperative morbidity and lack of functional benefit were identified as the most serious clinical shortcomings. Furthermore, the extent of regulatory policy in the use of stem cell therapies in China is unclear. In the absence of a valid clinical trials protocol, and more regulatory oversight, Western regulatory agencies advise patients and physicians to be cautious when selecting Chinese stem cell therapeutic centers.

Mexico

Stem cell treatment is currently being practiced at a clinical level in Mexico. An International Health Department Permit (COFEPRIS) is required.

This permit allows the use of stem cell types beyond those approved of in Western countries such as the United States or Europe. Stem cell therapies provided in Mexico utilize patient Adipose, Bone Marrow, or Donor Placenta sources.

South Korea

In 2005, South Korean scientists claimed to have generated stem cells that were tailored to match the recipient. Each of the 11 new stem cell lines was developed using somatic cell nuclear transfer (SCNT) technology. The resultant cells were thought to match the genetic material of the recipient, thus suggesting minimal to no cell rejection.

This study, however, was eventually discredited as the primary researcher,Dr. Woo Suk Hwang, admitted to using cells obtained from his research staff. In Dec 2005, claims were put forward that his research had been manipulated to wrongfully indicate positive results. Later that month, these claims were confirmed by an academic panel.

Ukraine

Today, Ukraine is permitted to perform clinical trials of stem cell treatments (Order of the MH of Ukraine [1] 630 "About carrying out clinical trials of stem cells", 2008) for the treatment of these pathologies: pancreatic necrosis, cirrhosis, hepatitis, burn disease, diabetes, multiple sclerosis, critical lower limb ischemia. The first medical institution granted the right to conduct clinical trials became the "Institute of Cell Therapy"(Kiev).

- Cardiovascular Cell Therapy Research Network (CCTRN)
- Fetal tissue implant
- Human Stem Cells Institute

Pharmacological

As of August 2008, several trials of various compounds are in development or ongoing, a few at the point of testing on larger numbers of people, known as phase III of clinical trials. Substances that have shown promise in initial experiments include dopamine receptor blockers, select dopamine antagonists, such as creatine and CoQ10, the antibiotic Minocycline, antioxidant-containing foods and nutrients, and antidepressants, including selective serotonin reuptake inhibitors such as sertraline, fluoxetine, and paroxetine. In 2010, minocycline was found to be ineffective for humans in a multi-center trial.

A 2009 study suggested strong association between specific single nucleotide polymorphisms alleles and CAG expansion also provides an opportunity of personalized therapeutics in HD where the clinical development of only a small number of allele-specific targets may be sufficient to treat up to 88% of the HD patient population.

Other

Recently it was shown that increasing BDNF expression may represent a useful strategy for Huntington's disease treatment.

- Gauchers disease is a genetic disease affecting metabolism. It is more treatable then most other genetic diseases, and can be treated with enzyme replacement therapy, medication miglustat, and bone marrow transplantion.

Classification

Gaucher's disease has three common clinical subtypes.

- Type I (or non-neuropathic type) is the most common form of the disease, occurring in approximately 1 in 50,000 live births. It occurs most often among persons of Ashkenazi Jewish heritage. Symptoms may begin early in life or in adulthood and include enlarged liver and grossly enlarged spleen (together hepatosplenomegaly); the spleen can rupture and cause additional complications. Skeletal weakness and bone disease may be extensive. Spleen enlargement and bone marrow replacement cause anemia, thrombocytopenia and leukopenia. The brain is not affected pathologically, but there may be lung and, rarely, kidney impairment. Patients in this group usually bruise easily (due to low levels of platelets) and experience fatigue due to low numbers of red blood cells. Depending on disease onset and severity, type 1 patients may live well into adulthood. Many patients have a mild form of the disease or may not show any symptoms.
- Type II (or acute infantile neuropathic Gaucher's disease) typically begins within 6 months of birth and has an incidence rate of approximately 1 in 100,000 live births. Symptoms include an enlarged liver and spleen, extensive and progressive brain damage, eye movement disorders, spasticity, seizures, limb rigidity, and a poor ability to suck and swallow. Affected children usually die by age 2.
- Type III (the chronic neuropathic form) can begin at any time in childhood or even in adulthood, and occurs in approximately 1 in 100,000 live births. It is characterized by slowly progressive but milder neurologic symptoms compared to the acute or type 2 version. Major symptoms include an enlarged spleen and/or liver, seizures, poor coordination, skeletal irregularities, eye movement disorders, blood disorders including anemia and respiratory problems. Patients often live into their early teen years and adulthood.

These subtypes have come under some criticism for not taking account of the full spectrum of observable symptoms (the phenotypes.) There are also compound heterozygous variations which considerably increase the complexity of predicting disease course.

Signs and Symptoms

- Painless hepatomegaly and splenomegaly: the size of the spleen can be 1500-3000 ml, as opposed to the normal size of 50-200 ml. Splenomegaly may decrease the affected individual's capacity for eating by exerting pressure on the stomach. While painless, enlargement of spleen increases the risk of splenic rupture.

- Hypersplenism and pancytopenia: the rapid and premature destruction of blood cells, leading to anemia, neutropenia, leukopenia, and thrombocytopenia (with an increased risk of infection and bleeding).
- Cirrhosis of the liver is rare
- Severe pain associated with joints and bones, frequently presenting in hips and knees.
- Neurological symptoms occur only in some types of Gaucher's:
  - o Type II: serious convulsions, hypertonia, mental retardation, apnea.
  - o Type III: muscle twitches known as myoclonus, convulsions, dementia, ocular muscle apraxia.
- Osteoporosis: 75% develop visible bony abnormalities due to the accumulated glucosylceramide. A deformity of the distal femur in the shape of an Erlenmeyer flask is commonly described (aseptic necrosis of the femur joint).
- Yellowish-brown skin pigmentation

Pathophysiology

The disease is caused by a defect in housekeeping gene lysosomal glucocerebrosidase (also known as beta-glucosidase, EC 3.2.1.45, PDB 1OGS) on the first chromosome (1q21). The enzyme is a 55.6 KD, 497 amino acids long protein that catalyses the breakdown of glucosylceramidase, a cell membrane constituent of red and white blood cells.

The macrophages that clear these cells are unable to eliminate the waste product, which accumulates in fibrils, and turn into Gaucher cells, which appear on light microscopy to resemble crumpled-up paper.

In the brain (type II and III), glucosylceramidase accumulates due to the turnover of complex lipids during brain development and the formation of the myelin sheath of nerves.

Different mutations in the beta-glucosidase determine the remaining activity of the enzyme, and, to a large extent, the phenotype.

Heterozygotes for particular acid beta-glucosidase mutations carry about a fivefold risk of developing Parkinson's disease, making this the most common known genetic risk-factor for Parkinson's. A study of 1525 Gaucher patients in the United States suggested that while cancer risk is not elevated, particular malignancies (non-Hodgkin lymphoma, melanoma and pancreatic cancer) occurred at a 2-3 times higher rate.

Genetics

The three types of Gaucher's disease are inherited in an autosomal recessive fashion. Both parents must be carriers in order for a child to be affected. If both parents are carriers, there is a one in four, or 25%, chance with each pregnancy for an affected child. Genetic counseling and genetic testing is recommended for families who may be carriers of mutations.

Each type has been linked to particular mutations. In all, there are about 80 known mutations, grouped into three main types:

- Type I (N370S homozygote), the most common, also called the "non-neuropathic" type occurs mainly in Ashkenazi Jews, at 100 times the occurrence in the general populace. The median age at diagnosis is 28 years of age, and life expectancy is mildly decreased. There are no neurological symptoms.
- Type II (1 or 2 alleles L444P) is characterized by neurological problems in small children. The enzyme is hardly released into the lysosomes. Prognosis is dismal: most die before reaching the third birthday.
- Type III (also 1-2 copies of L444P, possibly delayed by protective polymorphisms) occurs in Swedish patients from the Norrbotten region. This group develops the disease somewhat later, but most die before their 30th birthday.

Diaz et al. suggest that the Gaucher-causing mutations entered the Ashkenazi Jewish gene pool in the early Middle Ages (48-55 generations ago).

Diagnosis

Micrograph showing crinkled paper macrophages in the marrow space in a case of Gaucher disease. H&E stain.

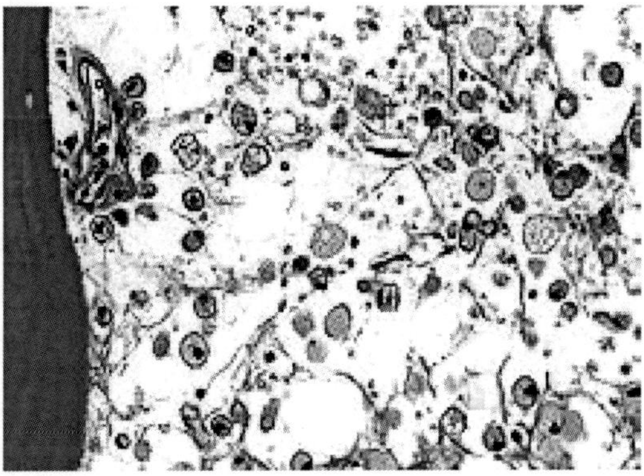

A definitive diagnosis is made with genetic testing. As there are numerous different mutations, sequencing of the beta-glucosidase gene is sometimes necessary to confirm the diagnosis. Prenatal diagnosis is available, and is useful when there is a known genetic risk factor. A diagnosis can also be implied by biochemical abnormalities such as high alkaline phosphatase, angiotensin-converting enzyme (ACE) and immunoglobulin levels, or by cell analysis showing "crinkled paper" cytoplasm and glycolipid-laden macrophages. Some lysosomal enzymes are elevated, including tartrate-resistant acid phosphatase, hexosaminidase, and a human chitinase, chitotriosidase. This latter enzyme has proved to be very useful for monitoring Gaucher's disease activity in response to treatment, and may reflect the severity of the disease

Treatment

For type 1 and most type 3 patients, enzyme replacement treatment with intravenous recombinant glucocerebrosidase (imiglucerase) can dramatically decrease liver and spleen size, reduce skeletal abnormalities, and reverse other manifestations. This treatment costs approximately $200,000 annually for a single patient and should be continued for life.

The rarity of the disease means that dose-finding studies have been difficult to conduct, so there remains controversy over the optimal dose and dosing frequency. Due to the low incidence, this has become an orphan drug in many countries, meaning that a government recognizes and accommodates the financial constraints that limit research into drugs that address a small population.Velaglucerase alfa was approved by the Food and Drug Administration (FDA) as an alternative treatment on February 26, 2010.

Successful bone marrow transplantation cures the non-neurological manifestations of the disease, because it introduces a monocyte population with active beta-glucosidase. However, this procedure carries significant risk and is rarely performed in Gaucher patients. Surgery to remove the spleen (splenectomy) may be required on rare occasions if the patient is anemic or when the enlarged organ affects the patient's comfort.

Blood transfusion may benefit some anemic patients. Other patients may require joint replacement surgery to improve mobility and quality of life. Other treatment options include antibiotics for infections, antiepileptics for seizures, bisphosphonates for bone lesions, and liver transplants. Substrate reduction therapy may prove to be effective in stopping Type 2, as it can cross through the blood barrier into the brain. There is currently no effective treatment for the severe brain damage that may occur in patients with types 2 and 3 Gaucher disease. Gene therapy may be a future step.

Gaucher's disease has recently become a target for more than one effort at pharmacological chaperoning, which involves the use of orally administered drugs that operate at a molecular level. Miglustat is one of these oral drugs. It was approved for the treatment of this disease in 2003. As of June 2009, another oral drug, isofagomine tartrate, is under development.

Epidemiology

The National Gaucher Foundation states that around 1 in 100 people in the general U.S. population is a carrier for type 1 Gaucher's disease, giving a prevalence of 1 in 40,000: among Ashkenazi Jews the rate of carriers is considerably higher, at roughly 1 in 15.

Type 2 Gaucher's disease shows no particular preference for any ethnic group.

Type 3 Gaucher's disease is especially common in the population of the Northern Swedish region of Norrbotten where the incidence of the disease is 1 in 50,000.

History

The disease was first recognised by the French doctor Philippe Gaucher, who originally described it in 1882 and lent his name to the condition. The biochemical basis for the disease would be elucidated in 1965. The first effective treatment for the disease, the drug Ceredase, was approved by the FDA in April 1991. An improved drug, Cerezyme, was approved by the FDA in May 1994 and has replaced the use of Ceredase.

Genetic epidemiology

Genetic epidemiology is the study of the role of genetic factors in determining health and disease in families and in populations, and the interplay of such genetic factors with environmental factors. In slightly more formal language, genetic epidemiology was defined by Morton as "a science which deals with the etiology, distribution, and control of disease in groups of relatives and with inherited causes of disease in populations". It is closely allied to both molecular epidemiology and statistical genetics, but these overlapping fields each have distinct emphases, societies and journals.

Traditionally, the study of the role of genetics in disease progresses through the following study designs, each answering a slightly different question:

- Familial aggregation studies: Is there a genetic component to the disease, and what are the relative contributions of genes and environment?
- Segregation studies: What is the pattern of inheritance of the disease (e.g. dominant or recessive)?
- Linkage studies: On which part of which chromosome is the disease gene located?
- Association studies: Which allele of which gene is associated with the disease?

This traditional approach has proved highly successful in identifying monogenic disorders and locating the genes responsible.

More recently, the scope of genetic epidemiology has expanded to include common diseases for which many genes each make a smaller contribution (polygenic, multifactorial or multigenic disorders). This has developed rapidly in the first decade of the 21st century following completion of the Human Genome Project, as advances in genotyping technology and associated reductions in cost has made it feasible to conduct large-scale genome-wide association studies that genotype many thousands of single nucleotide polymorphisms in thousands of individuals. These have led to the discovery of many genetic polymorphisms that influence the risk of developing many common diseases.

Genotoxicity Bioindicator

Cluster analysis in epidemiology is the method for identifying wide spread population damage (as mercury contamination in Minamata Bay) but a genotoxicity bioindicator is a preventive way of seeing congenital malformations epidemic in human populations.

A biological monitor, or biomonitor, is defined as an organism that provides quantitative information on the quality of the environment around it. Therefore, a good biomonitor will indicate the presence of the pollutant and also attempt to provide additional information about the amount and intensity of the exposure.

Endocrine Disruption (feminization); Reproductive Failure causes include early embryonic death, embryo toxicity and abnormal parental behaviour during incubation; Eggshell Thinning including Double-crested Cormorants, Ospreys, Bald Eagles, Black-crowned Night-Herons and Herring Gulls, maybe are caused by pollution or another genotoxic factor. Malformed frogs are caused by parasites, genetic disease or pollution? Crossed bills, jaw defects, extra limbs, and malformed feet, joints and eyes were found in Herring Gulls and at least eight other species of fish-eating waterbirds from great lakes, malformed frogs were found in Minnesota.

Inborn errors of metabolism

Inborn errors of metabolism comprise a large class of genetic diseases involving disorders of metabolism. The majority are due to defects of single genes that code for enzymes that facilitate conversion of various substances (substrates) into others (products). In most of the disorders, problems arise due to accumulation of substances which are toxic or interfere with normal function, or to the effects of reduced ability to synthesize essential compounds. Inborn errors of metabolism are now often referred to as congenital metabolic diseases or inherited metabolic diseases.

The term inborn error of metabolism was coined by a British physician, Archibald Garrod (1857–1936), in the early 20th century (1908). He is known for work that prefigured the "one gene-one enzyme" hypothesis, based on his studies on the nature and inheritance of alkaptonuria. His seminal text, Inborn Errors of Metabolism was published in 1923.

Major Categories of Inherited Metabolic Diseases

Traditionally the inherited metabolic diseases were categorized as disorders of carbohydrate metabolism, amino acid metabolism, organic acid metabolism, or lysosomal storage diseases. In recent decades, hundreds of new inherited disorders of metabolism have been discovered and the categories have proliferated. Following are some of the major classes of congenital metabolic diseases, with prominent examples of each class.

Many others do not fall into these categories. ICD-10 codes are provided where available:

- Disorders of carbohydrate metabolism
  - E.g., glycogen storage disease
- Disorders of amino acid metabolism
  - E.g., phenylketonuria, maple syrup urine disease, glutaric acidemia type 1
- Disorders of organic acid metabolism (organic acidurias)
  - E.g., alcaptonuria
- Disorders of fatty acid oxidation and mitochondrial metabolism
  - E.g., Medium-chain acyl-coenzyme A dehydrogenase deficiency (often shortened to MCADD.)
- Disorders of porphyrin metabolism
  - E.g., acute intermittent porphyria
- Disorders of purine or pyrimidine metabolism
  - E.g., Lesch-Nyhan syndrome
- Disorders of steroid metabolism
  - E.g., congenital adrenal hyperplasia
- Disorders of mitochondrial function
  - E.g ., Kearns-Sayre syndrome
- Disorders of peroxisomal function
  - E.g., Zellweger syndrome
- Lysosomal storage disorders
  - E.g., Gaucher's disease
  - E.g., Niemann Pick disease

Incidence

In a study in British Columbia, the overall incidence of the inborn errors of metabolism were estimated to be 70 per 100,000 live births or 1 in 1,400 births, overall representing more than approximately 15% of single gene disorders in the population.

Type of Inborn Error	*Incidence*	
Disease involving amino acids (e.g. PKU), organic acids, primary lactic acidosis, galactosemia, or a urea cycle disease	24 per 100 000 births	1 in 4,200
Lysosomal storage disease	8 per 100 000 births	1 in 12,500
Peroxisomal disorder	~3 to 4 per 100 000 of births	~1 in 30,000
Respiratory chain-based mitochondrial disease	~3 per 100 000 births	1 in 33,000
Glycogen storage disease	2.3 per 100 000 births	1 in 43,000

Manifestations and Presentations

Because of the enormous number of these diseases and wide range of systems affected, nearly every "presenting complaint" to a doctor may have a congenital metabolic disease as a possible cause, especially in childhood. The following are examples of potential manifestations affecting each of the major organ systems: many manifestations may develop

- Growth failure, failure to thrive, weight loss
- Ambiguous genitalia, delayed puberty, precocious puberty
- Developmental delay, seizures, dementia, encephalopathy, stroke
- Deafness, blindness, pain agnosia
- Skin rash, abnormal pigmentation, lack of pigmentation, excessive hair growth, lumps and bumps
- Dental abnormalities
- Immunodeficiency, thrombocytopenia, anemia, enlarged spleen, enlarged lymph nodes
- Many forms of cancer
- Recurrent vomiting, diarrhea, abdominal pain
- Excessive urination, renal failure, dehydration, edema
- Hypotension, heart failure, enlarged heart, hypertension, myocardial infarction

- Hepatomegaly, jaundice, liver failure
- Unusual facial features, congenital malformations
- Excessive breathing (hyperventilation), respiratory failure
- Abnormal behavior, depression, psychosis
- Joint pain, muscle weakness, cramps
- Hypothyroidism, adrenal insufficiency, hypogonadism, diabetes mellitus

Diagnostic Techniques

Dozens of congenital metabolic diseases are now detectable by newborn screening tests, especially the expanded testing using mass spectrometry.

This is an increasingly common way for the diagnosis to be made and sometimes results in earlier treatment and a better outcome. There is a revolutionary GC/MS based technology with an integrated analytics system, which has now made it possible to test a newborn for over 100 genetic metabolic disorders.

Because of the multiplicity of conditions, many different diagnostic tests are used for screening. An abnormal result is often followed by a subsequent "definitive test" to confirm the suspected diagnosis.

Common screening tests used in the last sixty years:

- Ferric chloride test (turned colors in reaction to various abnormal metabolites in urine)
- Ninhydrin paper chromatography (detected abnormal amino acid patterns)
- Guthrie bacterial inhibition assay (detected a few amino acids in excessive amounts in blood) The dried blood spot can be used for multianalyte testing using Tandem Mass Spectrometry (MS/MS). This given an indication for a disorder. The same has to be further confirmed by enzyme assays, GC/MS or DNA Testing.
- Quantitative measurement of amino acids in plasma and urine
- Urine organic acid analysis by Gas chromatography-mass spectrometry
- Plasma acylcarnitines analysis by mass spectrometry
- Urine purines and pyrimidines analysis by Gas chromatography-mass spectrometry

Specific diagnostic tests (or focused screening for a small set of disorders):

- Tissue biopsy or necropsy: liver, muscle, brain, bone marrow
- Skin biopsy and fibroblast cultivation for specific enzyme testing
- Specific DNA testing

Treatment

In the middle of the 20th century the principal treatment for some of the amino acid disorders was restriction of dietary protein and all other care was simply management of complications. In the past twenty years, enzyme replacement, gene transfer, and organ transplantation have become available and beneficial for many previously untreatable disorders. Some of the more common or promising therapies are listed:

- Dietary restriction
  - o E.g., reduction of dietary protein remains a mainstay of treatment for phenylketonuria and other amino acid disorders
- Dietary supplementation or replacement
  - o E.g., oral ingestion of cornstarch several times a day helps prevent people with glycogen storage diseases from becoming seriously hypoglycemic.
- Vitamins
  - o E.g., thiamine supplementation benefits several types of disorders that cause lactic acidosis.
- Intermediary metabolites, compounds, or drugs that facilitate or retard specific metabolic pathways
- Dialysis
- Enzyme replacement E.g. Acid-alpha glucosidase for Pompe disease
- Gene transfer
- Bone marrow or organ transplantation
- Treatment of symptoms and complications
- Prenatal diagnosis and avoidance of pregnancy or abortion of an affected fetus

Resources

For clinicians and scientists in the field of inborn errors of metabolism, good resources include books by Scriver, Fernandes, Clarke, Blau (diagnosis), Blau (treatment), Lyon, Nyhan, Hoffmann, and Zschocke. Other resources include genetests, orphanet, OMIM, Metab-L, societies such as the SSIEM, the SIMD and links therein. For medical students and clinicians looking for overviews of the field, such reviews can be found on pubmed and in good pediatric textbooks (e.g. articles by Saudubray, Ellaway, Raghuveer or Burton, and textbooks by Hay or Behrman).

For patients, their families or other individuals seeking good information and support groups, the National Institutes of Health offers the office of rare diseases, genetics home reference, medlineplus and health information. The National Human Genome Research Institute hosts an information center, a section for patients and the public and additional educational resources. Support groups can be found at NORD, Genetic Alliance and Orphanet. The genetic education center at the KUMC has many more useful links.

Medical Genetics

Medical genetics is the specialty of medicine that involves the diagnosis and management of hereditary disorders. Medical genetics differs from Human genetics in that human genetics is a field of scientific research that may or may not apply to medicine, but medical genetics refers to the application of genetics to medical care.

For example, research on the causes and inheritance of genetic disorders would be considered within both human genetics and medical genetics, while the diagnosis, management, and counseling of individuals with genetic disorders would be considered part of medical genetics.

In contrast, the study of typically non-medical phenotypes such as the genetics of eye color would be considered part of human genetics, but not necessarily relevant to medical genetics (except in situations such as albinism). Genetic medicine is a newer term for medical genetics and incorporates areas such as gene therapy, personalized medicine, and the rapidly emerging new medical specialty, predictive medicine.

Scope

Medical genetics encompasses many different areas, including clinical practice of physicians, genetic counselors, and nutritionists, clinical diagnostic laboratory activities, and research into the causes and inheritance of genetic disorders. Examples of conditions that fall within the scope of medical genetics include birth defects and dysmorphology, mental retardation, autism, and mitochondrial disorders, skeletal dysplasia, connective tissue disorders, cancer genetics, teratogens, and prenatal diagnosis.

Medical genetics is increasingly becoming relevant to many common diseases. Overlaps with other medical specialties are beginning to emerge, as recent advances in genetics are revealing etiologies for neurologic, endocrine, cardiovascular, pulmonary, ophthalmologic, renal, psychiatric, and dermatologic conditions.

Subspecialties

In some ways, many of the individual fields within medical genetics are hybrids between clinical care and research. This is due in part to recent advances in science and technology (for example, the Human genome project) that have enabled an unprecedented understanding of genetic disorders.

Clinical Genetics

Clinical genetics is the practice of clinical medicine with particular attention to hereditary disorders. Referrals are made to genetics clinics for a variety of reasons, including birth defects, developmental delay, autism, epilepsy, short stature, and many others. Examples of genetic syndromes that are commonly seen in the genetics clinic include chromosomal rearrangements, Down syndrome, DiGeorge syndrome (22q11.2 Deletion Syndrome), Fragile X syndrome, Marfan syndrome, Neurofibromatosis, Turner syndrome, and Williams syndrome.

Metabolic/Biochemical Genetics

Metabolic (or biochemical) genetics involves the diagnosis and management of inborn errors of metabolism in which patients have enzymatic deficiencies that perturb biochemical pathways involved in metabolism of carbohydrates, amino acids, and lipids. Examples of metabolic disorders include galactosemia, glycogen storage disease, lysosomal storage disorders, metabolic acidosis, peroxisomal disorders, phenylketonuria, and urea cycle disorders.

Cytogenetics

Cytogenetics is the study of chromosomes and chromosome abnormalities. While cytogenetics historically relied on microscopy to analyze chromosomes, new molecular technologies such as array comparative genomic hybridization are now becoming widely used. Examples of chromosome abnormalities include aneuploidy, chromosomal rearrangements, and genomic deletion/duplication disorders.

Molecular Genetics

Molecular genetics involves the discovery of and laboratory testing for DNA mutations that underlie many single gene disorders. Examples of single gene disorders include achondroplasia, cystic fibrosis, Duchenne muscular dystrophy, hereditary breast cancer (BRCA1/2), Huntington disease, Marfan syndrome, Noonan syndrome, and Rett syndrome. Molecular tests are also used in the diagnosis of syndromes involving epigenetic abnormalities, such as Angelman syndrome, Beckwith-Wiedemann syndrome, Prader-willi syndrome, and uniparental disomy.

Mitochondrial genetics

Mitochondrial genetics concerns the diagnosis and management of mitochondrial disorders, which have a molecular basis but often result in biochemical abnormalities due to deficient energy production. There exists some overlap between medical genetic diagnostic laboratories and molecular pathology.

Genetic Counseling

Genetic counseling is the process of providing information about genetic conditions, diagnostic testing, and risks in other family members, within the framework of nondirective counseling. Genetic counselors are non-physician members of the medical genetics team who specialize in family risk assessment and counseling of patients regarding genetic disorders. The precise role of the genetic counselor varies somewhat depending on the disorder.

History

Although genetics has its roots back in the 19th century with the work of the Bohemian monk Gregor Mendel and other pioneering scientists, human genetics emerged later. It started to develop, albeit slowly, during the first half of the 20th century. Mendelian (single-gene) inheritance was studied in a number of important disorders such as albinism, brachydactyly (short fingers and toes), and hemophilia. Mathematical approaches were also devised and applied to human genetics. Population genetics was created.

Medical genetics was a late developer, emerging largely after the close of World War II (1945) when the eugenics movement had fallen into disrepute. The Nazi misuse of eugenics sounded its death knell. Shorn of eugenics, a scientific approach could be used and was applied to human and medical genetics. Medical genetics saw an increasingly rapid rise in the second half of the 20th century and continues in the 21st century.

Notable Practitioners

- Victor McKusick (1921–2008), "The Father of Genetic Medicine" and principal author of Mendelian Inheritance in Man.

Current Practice

The clinical setting in which patients are evaluated determines the scope of practice, diagnostic, and therapeutic interventions. For the purposes of general discussion, the typical encounters between patients and genetic practitioners may involve:

- Referral to an out-patient genetics clinic (pediatric, adult, or combined) or an in-hospital consultation, most often for diagnostic evaluation.
- Specialty genetics clinics focusing on management of inborn errors of metabolism, skeletal dysplasia, or lysosomal storage diseases.
- Referral for counseling in a prenatal genetics clinic to discuss risks to the pregnancy (advanced maternal age, teratogen exposure, family history of a genetic disease), test results (abnormal maternal serum screen, abnormal ultrasound), and/or options for prenatal diagnosis (typically amniocentesis or chorionic villus sampling).

- Multidisciplinary specialty clinics that include a clinical geneticist or genetic counselor (cancer genetics, cardiovascular genetics, craniofacial or cleft lip/palate, hearing loss clinics, muscular dystrophy/ neurodegenerative disorder clinics).

Diagnostic Evaluation

Each patient will undergo a diagnostic evaluation tailored to their own particular presenting signs and symptoms. The geneticist will establish a differential diagnosis and recommend appropriate testing. Increasingly, clinicians use SimulConsult, paired with the National Library of Medicine Gene Review articles, to narrow the list of hypotheses (known as the differential diagnosis) and identify the tests that are relevant for a particular patient. These tests might evaluate for chromosomal disorders, inborn errors of metabolism, or single gene disorders.

Chromosome Studies

Chromosome studies are used in the general genetics clinic to determine a cause for developmental delay/mental retardation, birth defects, dysmorphic features, and/or autism. Chromosome analysis is also performed in the prenatal setting to determine whether a fetus is affected with aneuploidy or other chromosome rearrangements. Finally, chromosome abnormalities are often detected in cancer samples. A large number of different methods have been developed for chromosome analysis:

- Chromosome analysis using a karyotype involves special stains that generate light and dark bands, allowing identification of each chromosome under a microscope.
- Fluorescence in situ hybridization (FISH) involves fluorescent labeling of probes that bind to specific DNA sequences, used for identifying aneuploidy, genomic deletions or duplications, characterizing chromosomal translocations and determining the origin of ring chromosomes.
- Chromosome painting is a technique that uses fluorescent probes specific for each chromosome to differentially label each chromosome. This technique is more often used in cancer cytogenetics, where complex chromosome rearrangements can occur.
- Array comparative genomic hybridization is a new molecular technique that involves hybridization of an individual DNA sample to a glass slide or microarray chip containing molecular probes (ranging from large ~200kb bacterial artificial chromosomes to small oligonucleotides) that represent unique regions of the genome. This method is particularly sensitive for detection of genomic gains or losses across the genome but

does not detect balanced translocations or distinguish the location of duplicated genetic material (for example, a tandem duplication versus an insertional duplication).

Basic Metabolic Studies

Biochemical studies are performed to screen for imbalances of metabolites in the bodily fluid, usually the blood (plasma/serum) or urine, but also in cerebrospinal fluid (CSF). Specific tests of enzyme function (either in leukocytes, skin fibroblasts, liver, or muscle) are also employed under certain circumstances. In the US, the newborn screen incorporates biochemical tests to screen for treatable conditions such as galactosemia and phenylketonuria (PKU).

Patients suspected to have a metabolic condition might undergo the following tests:

- Quantitative amino acid analysis is typically performed using the ninhydrin reaction, followed by liquid chromatography to measure the amount of amino acid in the sample (either urine, plasma/serum, or CSF). Measurement of amino acids in plasma or serum is used in the evaluation of disorders of amino acid metabolism such as urea cycle disorders, maple syrup urine disease, and PKU. Measurement of amino acids in urine can be useful in the diagnosis of cystinuria or renal Fanconi syndrome as can be seen in cystinosis.
- Urine organic acid analysis can be either performed using quantitative or qualitative methods, but in either case the test is used to detect the excretion of abnormal organic acids. These compounds are normally produced during bodily metabolism of amino acids and odd-chain fatty acids, but accumulate in patients with certain metabolic conditions.
- The acylcarnitine combination profile detects compounds such as organic acids and fatty acids conjugated to carnitine. The test is used for detection of disorders involving fatty acid metabolism, including MCAD.
- Pyruvate and lactate are byproducts of normal metabolism, particularly during anaerobic metabolism. These compounds normally accumulate during exercise or ischemia, but are also elevated in patients with disorders of pyruvate metabolism or mitochondrial disorders.
- Ammonia is an end product of amino acid metabolism and is converted in the liver to urea through a series of enzymatic reactions termed the urea cycle. Elevated ammonia can therefore be detected in patients with urea cycle disorders, as well as other conditions involving liver failure.
- Enzyme testing is performed for a wide range of metabolic disorders to confirm a diagnosis suspected based on screening tests.

Molecular Studies

- DNA sequencing is used to directly analyze the genomic DNA sequence of a particular gene. In general, only the parts of the gene that code for the expressed protein (exons) and small amounts of the flanking untranslated regions and introns are analyzed. Therefore, although these tests are highly specific and sensitive, they do not routinely identify all of the mutations that could cause disease.
- DNA methylation analysis is used to diagnose certain genetic disorders that are caused by disruptions of epigenetic mechanisms such as genomic imprinting and uniparental disomy.
- Southern blotting is an early technique basic on detection of fragments of DNA separated by size through gel electrophoresis and detected using radiolabeled probes. This test was routinely used to detect deletions or duplications in conditions such as Duchenne muscular dystrophy but is being replaced by high-resolution array comparative genomic hybridization techniques. Southern blotting is still useful in the diagnosis of disorders caused by trinucleotide repeats.
- Short tandem repeats are unique markers that can be used to determine haplotypes and are used in identity testing for maternal cell contamination.

Treatments

Each cell of the body contains the hereditary information (DNA) wrapped up in structures called chromosomes. Since genetic syndromes are typically the result of alterations of the chromosomes or genes, there is no treatment currently available that can correct the genetic alterations in every cell of the body. Therefore, there is currently no "cure" for genetic disorders. However, for many genetic syndromes there is treatment available to manage the symptoms. In some cases, particularly inborn errors of metabolism, the mechanism of disease is well understood and offers the potential for dietary and medical management to prevent or reduce the long-term complications. In other cases, infusion therapy is used to replace the missing enzyme. Current research is actively seeking to use gene therapy or other new medications to treat specific genetic disorders.

Management of Metabolic Disorders

In general, metabolic disorders arise from enzyme deficiencies that disrupt normal metabolic pathways. Compound "A" is metabolized to "B" by enzyme "X", compound "B" is metabolized to "C" by enzyme "Y", and compound "C" is metabolized to "D" by enzyme "Z". If enzyme "Z" is missing, compound "D" will be missing, while compounds "A", "B", and "C" will build up. The pathogenesis of this particular condition could result from lack of compound "D", if it is critical for some cellular function, or from toxicity due to excess "A", "B", and/or "C". Treatment of the metabolic disorder could be achieved through dietary supplementation of compound "D" and dietary restriction of compounds "A", "B", and/or "C" or by treatment with a medication that promoted disposal of excess "A", "B", or "C". Another approach that can be taken is enzyme replacement therapy, in which a patient is given an infusion of the missing enzyme.

Diet

Dietary restriction and supplementation are key measures taken in several well-known metabolic disorders, including galactosemia, phenylketonuria (PKU), maple syrup urine disease, organic acidurias and urea cycle disorders. Such restrictive diets can be difficult for the patient and family to maintain, and require close consultation with a nutritionist who has special experience in metabolic disorders. The composition of the diet will change depending on the caloric needs of the growing child and special attention is needed during a pregnancy if a woman is affected with one of these disorders.

Medication

Medical approaches include enhancement of residual enzyme activity (in cases where the enzyme is made but is not functioning properly), inhibition of other enzymes in the biochemical pathway to prevent buildup of a toxic compound, or diversion of a toxic compound to another form that can be excreted. Examples include the use of high doses of pyridoxine (vitamin B6) in some patients with homocystinuria to boost the activity of the residual cystathione synthase enzyme, administration of biotin to restore activity of several enzymes affected by deficiency of biotinidase, treatment with NTBC in Tyrosinemia to inhibit the production of succinylacetone which causes liver toxicity, and the use of sodium benzoate to decrease ammonia build-up in urea cycle disorders.

Enzyme Replacement Therapy

Certain lysosomal storage diseases are treated with infusions of a recombinant enzyme (produced in a laboratory), which can reduce the accumulation of the compounds in various tissues. Examples include Gaucher disease, Fabry disease, Mucopolysaccharidoses and Glycogen storage disease type II. Such treatments are limited by the ability of the enzyme to reach the affected areas (the blood brain barrier prevents enzyme from reaching the brain, for example), and can sometimes be associated with allergic reactions. The long-term clinical effectiveness of enzyme replacement therapies vary widely among different disorders.

Other Examples

- Angiotensin receptor blockers in Marfan syndrome & Loeys-Dietz
- Bone marrow transplantation
- Gene therapy

Ethical, Legal and Social Implications

Genetic information provides a unique type of knowledge about an individual and his/her family, fundamentally different than a typically laboratory test that provides a "snapshot" of an individual's health status. The unique status of genetic information and inherited disease has a number of ramifications with regard to ethical, legal, and societal concerns.

Societies

The more empirical approach to human and medical genetics was formalized by the founding in 1948 of the American Society of Human Genetics. The Society first began annual meetings that year (1948) and its international counterpart, the International Congress of Human Genetics, has met every 5 years since its inception in 1956. The Society publishes the American Journal of Human Genetics on a monthly basis.

Medical genetics is now recognized as a distinct medical specialty in the U.S. with its own approved board (the American Board of Medical Genetics) and clinical specialty college (the American College of Medical Genetics). The College holds an annual scientific meeting, publishes a monthly journal, Genetics in Medicine, and issues position papers and clinical practice guidelines on a variety of topics relevant to human genetics.

Research

This section requires expansion with:

The broad range of research in medical genetics reflects the overall scope of this field, including basic research on genetic inheritance and the human genome, mechanisms of genetic and metabolic disorders, translational research on new treatment modalities, and the impact of genetic testing

Basic Genetics Research

Basic research geneticists usually undertake research in universities, biotechnology firms and research institutes.

Allelic Architecture of Disease

Sometimes the link between a disease and an unusual gene variant is more subtle. The genetic architecture of common diseases is an important factor in determining the extent to which patterns of genetic variation influence group differences in health outcomes. According to the common disease/common variant hypothesis, common variants present in the ancestral population before the dispersal of modern humans from Africa play an important role in human diseases. Genetic variants associated with Alzheimer disease, deep venous thrombosis, Crohn disease, and type 2 diabetes appear to adhere to this model. However, the generality of the model has not yet been established and, in some cases, is in doubt. Some diseases, such as many common cancers, appear not to be well described by the common disease/common variant model.

Another possibility is that common diseases arise in part through the action of combinations of variants that are individually rare. Most of the disease-associated alleles discovered to date have been rare, and rare variants are more likely than common variants to be differentially distributed among groups distinguished by ancestry. However, groups could harbor different, though perhaps overlapping, sets of rare variants, which would reduce contrasts between groups in the incidence of the disease.

The number of variants contributing to a disease and the interactions among those variants also could influence the distribution of diseases among groups. The difficulty that has been encountered in finding contributory alleles for complex diseases and in replicating positive associations suggests that many complex diseases involve numerous variants rather than a moderate number of alleles, and the influence of any given variant may depend in critical ways on the genetic and environmental background. If many alleles are required to increase susceptibility to a disease, the odds are low that the necessary combination of alleles would become concentrated in a particular group purely through drift.

Population Substructure in Genetics Research

One area in which population categories can be important considerations in genetics research is in controlling for confounding between population substructure, environmental exposures, and health outcomes.

Association studies can produce spurious results if cases and controls have differing allele frequencies for genes that are not related to the disease being studied, although the magnitude of this problem in genetic association studies is subject to debate. Various methods have been developed to detect and account for population substructure, but these methods can be difficult to apply in practice.

Population substructure also can be used to advantage in genetic association studies. For example, populations that represent recent mixtures of geographically separated ancestral groups can exhibit longer-range linkage disequilibrium between susceptibility alleles and genetic markers than is the case for other populations. Genetic studies can use this admixture linkage disequilibrium to search for disease alleles with fewer markers than would be needed otherwise. Association studies also can take advantage of the contrasting experiences of racial or ethnic groups, including migrant groups, to search for interactions between particular alleles and environmental factors that might influence health.

Human Genetic Clustering

Human genetic clustering analysis uses mathematical cluster analysis of the degree of similarity of genetic data between individuals and groups to infer population structures and assign individuals to groups that often correspond with their self-identified geographical ancestry. A similar analysis can be done using principal components analysis, which in earlier research was a popular method. Many of recent studies in the past few years have returned to using principal components analysis.

Studies

Phylogenetic tree by Cavalli-Sforza & al. (1994)

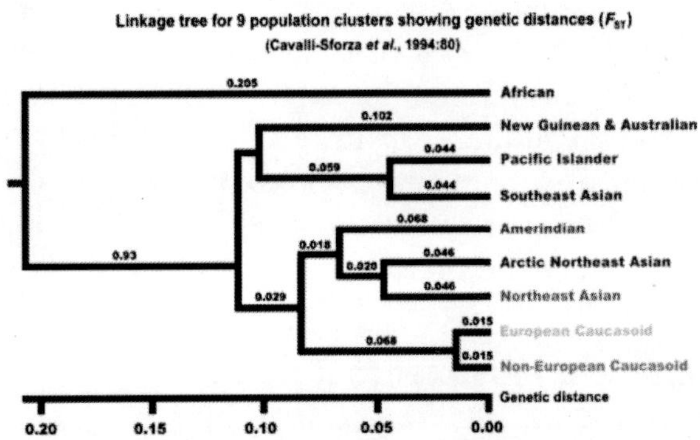

F_{ST} distance matrix for the 9 clusters shown above (x10,000 with standard errors obtained by bootstrap analysis)

	AFR	NEC	EUC	NEA	ANE	AME	SEA	PAI	NGA
African	0.0								
Non-European Caucasian	1340.0 ± 301	0.0							
European Caucasian	1655.6 ± 416	154.7 ± 29	0.0						
Northeast Asian	1979.1 ± 452	640.4 ± 134	938.2 ± 217	0.0					
Arctic North-east Asian	2008.5 ± 387	708.2 ± 160	746.7 ± 210	459.7 ± 98	0.0				
Amerindian	2261.4 ± 434	955.5 ± 204	1038.2 ± 276	746.5 ± 183	577.4 ± 89	0.0			
Southeast Asian	2206.3 ± 529	939.6 ± 262	1240.4 ± 339	630.5 ± 299	1039.4 ± 326	1341.7 ± 418	0.0		
Pacific Islander	2505.4 ± 648	953.7 ± 230	1344.7 ± 354	723.8 ± 262	1181.2 ± 331	1740.7 ± 544	436.7 ± 87	0.0	
New Guinean and Australian	2472.0 ± 536	1179.1 ± 189	1345.7 ± 231	734.4 ± 118	1012.5 ± 257	1457.9 ± 283	1237.9 ± 277	808.7 ± 264	0.0

Neighbor-joining method, by Naruya Saitou and Masatoshi Nei (2002)

In 2004, Lynn Jorde and Steven Wooding argued that "Analysis of many loci now yields reasonably accurate estimates of genetic similarity among individuals, rather than populations. Clustering of individuals is correlated with geographic origin or ancestry."

Gene clusters from Rosenberg (2006) for K=7 clusters. (Cluster analysis divides a dataset into any prespecified number of clusters.) Individuals have genes from multiple clusters. The cluster prevalent only among the Kalash people (yellow) only splits off at K=7 and greater.

Human population structure can be inferred from multilocus DNA sequence data (Rosenberg et al. 2002, 2005). Individuals from 52 populations were examined at 993 DNA markers. This data was used to partition individuals into K = 2, 3, 4, 5, or 6 gene clusters. In this figure, the average fractional membership of individuals from each population is represented by horizontal bars partitioned into K colored segments.

A study by Neil Risch in 2005 used 326 microsatellite markers and self-identified race/ethnic group (SIRE), white (European American), African-American (black), Asian and Hispanic (individuals involved in the study had to choose from one of these categories), to representing discrete "populations", and showed distinct and non-overlapping clustering of the white,

African-American and Asian samples. The results were claimed to confirm the integrity of self-described ancestry: "We have shown a nearly perfect correspondence between genetic cluster and SIRE for major ethnic groups living in the United States, with a discrepancy rate of only 0.14%."(Tang, 2005)

Studies such as those by Risch and Rosenberg use a computer program called STRUCTURE to find human populations (gene clusters). It is a statistical program that works by placing individuals into one of two clusters based on their overall genetic similarity, many possible pairs of clusters are tested per individual to generate multiple clusters. These populations are based on multiple genetic markers that are often shared between different human populations even over large geographic ranges. The notion of a genetic cluster is that people within the cluster share on average similar allele frequencies to each other than to those in other clusters. (A. W. F. Edwards, 2003 but also infobox "Multi Locus Allele Clusters") In a test of idealised populations, the computer programme STRUCTURE was found to consistently under-estimate the numbers of populations in the data set when high migration rates between populations and slow mutation rates (such as single-nucleotide polymorphisms) were considered.

Nevertheless the Rosenberg et al. (2002) paper shows that individuals can be assigned to specific clusters to a high degree of accuracy. One of the underlying questions regarding the distribution of human genetic diversity is related to the degree to which genes are shared between the observed clusters. It has been observed repeatedly that the majority of variation observed in the global human population is found within populations. This variation is usually calculated using Sewall Wright's Fixation index (F_{ST}), which is an estimate of between to within group variation. The degree of human genetic variation is a little different depending upon the gene type studied, but in general it is common to claim that ~85% of genetic variation is found within groups, ~6–10% between groups within the same continent and ~6–10% is found between continental groups.

For example The Human Genome Project states "two random individuals from any one group are almost as different [genetically] as any two random individuals from the entire world." Sarich and Miele, however, have argued that estimates of genetic difference between individuals of different populations fail to take into account human diploidity.

The point is that we are diploid organisms, getting one set of chromosomes from one parent and a second from the other. To the extent that your mother and father are not especially closely related, then, those two sets of chromosomes will come close to being a random sample of the chromosomes in your population. And the sets present in some randomly chosen member of yours will also be about as different from your two sets as they are from one another. So how much of the variability will be distributed where? First is the 15 percent that is interpopulational.

The other 85 percent will then split half and half (42.5 percent) between the intra- and interindividual within-population comparisons. The increase in variability in between-population comparisons is thus 15 percent against the 42.5 percent that is between-individual within-population. Thus, 15/42.5 is 32.5 percent, a much more impressive and, more important, more legitimate value than 15 percent.

Additionally, Edwards (2003) claims in his essay "Lewontin's Fallacy" that: "It is not true, as Nature claimed, that 'two random individuals from any one group are almost as different as any two random individuals from the entire world'" and Risch et al. (2002) state "Two Caucasians are more similar to each other genetically than a Caucasian and an Asian." It should be noted that these statements are not the same. Risch et al.

Simply state that two indigenous individuals from the same geographical region are more similar to each other than either is to an indigenous individual from a different geographical region, a claim few would argue with. Jorde et al. put it like this:

The picture that begins to emerge from this and other analyses of human genetic variation is that variation tends to be geographically structured, such that most individuals from the same geographic region will be more similar to one another than to individuals from a distant region.

Whereas Edwards claims that it is not true that the differences between individuals from different geographical regions represent only a small proportion of the variation within the human population (he claims that within group differences between individuals are not almost as large as between group differences). Bamshad et al. (2004) used the data from Rosenberg et al.

(2002) to investigate the extent of genetic differences between individuals within continental groups relative to genetic differences between individuals between continental groups. They found that though these individuals could be classified very accurately to continental clusters, there was a significant degree of genetic overlap on the individual level, to the extent that, using 377 loci, individual Europeans were about 38% of the time more genetically similar to East Asians than to other Europeans.

The existence of allelic clines and the observation that the bulk of human variation is continuously distributed, has led some scientists to conclude that any categorization schema attempting to partition that variation meaningfully will necessarily create artificial truncations. (Kittles & Weiss 2003).

It is for this reason, Reanne Frank argues, that attempts to allocate individuals into ancestry groupings based on genetic information have yielded varying results that are highly dependent on methodological design. Serre and Pääbo (2004) make a similar claim:

The absence of strong continental clustering in the human gene pool is of practical importance. It has recently been claimed that "the greatest genetic structure that exists in the human population occurs at the racial level" (Risch et al. 2002). Our results show that this is not the case, and we see no reason to assume that "races" represent any units of relevance for understanding human genetic history.

In a response to Serre and Pääbo (2004), Rosenberg et al. (2005) make three relevant observations. Firstly they maintain that their clustering analysis is robust. Secondly they agree with Serre and Pääbo that membership of multiple clusters can be interpreted as evidence for clinality (isolation by distance), though they also comment that this may also be due to admixture between neighbouring groups (small island model). Thirdly they comment that evidence of clusterdness is not evidence for any concepts of "biological race".

Risch et al. (2002) state that "two Caucasians are more similar to each other genetically than a Caucasian and an Asian", but Bamshad et al. (2004) used the same data set as Rosenberg et al. (2002) to show that Europeans are more similar to Asians 38% of the time than they are to other Europeans when only 377 microsatellite markers are analysed. If a landmass is considered with variation distributed in one dimension (west-east). Top: Distribution of genetic variation if a small island model is considered; there are two "populations" with a narrow region of hybridisation where migration occurs. This pattern is "clustered".

Bottom: Distribution of genetic variation if isolation by distance is considered; all variation is gradual over the extent of the landmass. This pattern is "clinal".

Percentage similarity between two individuals from different clusters when 377 microsatellite markers are considered.

x	***Africans***	***Europeans***	***Asians***
Europeans	36.5	—	—
Asians	35.5	38.3	—
Indigenous Americans	26.1	33.4	35

In agreement with the observation of Bamshad et al. (2004), Witherspoon et al. (2007) have shown that many more than 326 or 377 microsatellite loci are required in order to show that individuals are always more similar to individuals in their own population group than to individuals in different population groups, even for three distinct populations.

Witherspoon et al. (2007) have argued that even when individuals can be reliably assigned to specific population groups, it may still be possible for two randomly chosen individuals from different populations/clusters to be more similar to each other than to a randomly chosen member of their own cluster.

They found that many thousands of genetic markers had to be used in order for the answer to the question "How often is a pair of individuals from one population genetically more dissimilar than two individuals chosen from two different populations?" to be "never". This assumed three population groups separated by large geographic ranges (European, African and East Asian).

The entire world population is much more complex and studying an increasing number of groups would require an increasing number of markers for the same answer.

Witherspoon et al. conclude that "caution should be used when using geographic or genetic ancestry to make inferences about individual phenotypes."

Clustering does not particularly correspond to continental divisions. Depending on the parameters given to their analytical program, Rosenberg and Pritchard were able to construct between divisions of between 4 and 20 clusters of the genomes studied, although they excluded analysis with more than 6 clusters from their published article. Probability values for various cluster configurations varied widely, with the single most likely configuration coming with 16 clusters although other 16-cluster configurations had low probabilities. Overall, "there is no clear evidence that K=6 was the best estimate" according to geneticist Deborah Bolnick (2008:76-77).

Population Groups in Biomedicine

Race and Health

Race and health research, often done in the United States, has found both current and historical racial differences in the frequency, treatments, and availability of treatments for several diseases. This can add up to significant group differences in variables such as life expectancy.

Many explanations for such differences have been argued, including socioeconomic factors (e.g., education, employment, and income), lifestyle behaviors (e.g., physical activity and alcohol intake), social environment (e.g., educational and economic opportunities, racial/ethnic discrimination, and neighborhood and work conditions), and access to preventive health-care services (e.g., cancer screening and vaccination) as well as to treatment (through lack of insurance, lack of hospitals in certain areas, etc.), among other environmental differences. Some diseases may also be influenced by genes which differ in frequency between groups, such as sickle-cell anemia, which occurs overwhelmingly among some black populations, although the significance in clinical medicine of race categories as a proxy for exact genotypes of individuals has been questioned.

Health Disparities

Health disparities refer to gaps in the quality of health and health care across racial and ethnic groups. The US Health Resources and Services Administration defines health disparities as "population-specific differences in the presence of disease, health outcomes, or access to health care." Health is measured through variables such as life expectancy and incidence of diseases.

Single Gene Disorders

There are many single gene genetic disorders that differ in frequency between different popultions. Some examples are discussed below.

Cystic fibrosis is the most common life-limiting autosomal recessive disease among people of European heritage.

Sickle-cell anemia is most prevalent in populations of sub-Saharan African ancestry, but it is also common among Latin-American, Indian, Saudi Arab, and Mediterranean populations such as Turkey, Greece, and Italy. Tay-Sachs Disease is an autosomal recessive disorder that is more frequent among Ashkenazi Jews than among other Jewish groups and non-Jewish populations.

Hereditary hemochromatosis is most common among those of Northern European ancestry, in particular those of Celtic descendent. Lactose intolerance is another examples of single gene genetic disorders that differ in frequency between populations.

Multifactorial Polygenic Diseases

These are many diseases which differ in frequency between different populations.

Disease Progression

Groups may differ in how a disease progresses.

Treatment

Groups may differ in their responses to treatment.

Medicines Targeted at Specific Races

Race-based medicine is the term for medicines that are targeted at specific ethnic clusters which are shown to have a propensity for a certain disorder.

The first example of this in the U.S. was when BiDil, a medication for congestive heart failure, was licensed specifically for use in American patients that self-identify as black. Previous studies had shown that African American patients with congestive heart failure generally respond less effectively to traditional treatments than white patients with similar conditions.

After two trials, BiDil was licensed exclusively for use in African American patients. Critics have argued that this particular licensing was unwarranted, since the trials did not in fact show that the drug was more effective in African Americans than in other groups, but merely that it was more effective in African Americans thanother similar drug.

It was also only tested in African American males, but not in any other racial groups or among women. This peculiar trial and licensing procedure has prompted suggestions that the licensing was in fact used as a race based advertising scheme.

Environmental Explanations

In multiracial societies such as the United States, racial groups differ greatly in regard to social and cultural factors such as socioeconomic status, healthcare, diet, and education.

There is also the presence of racism which some see as a very important explaining factor. Some argue that for many diseases racial differences would disappear if all environmental factors could be controlled for. Article about race and health in the United States for a discussion of such factors. These factors may or may not be appropriate in other nations.

Ethnic minorities may also have specific health care needs which need to be taken into consideration by health services in order to tackle health disparities.

Genetic Explanations

Evolutionary Explanations

Genes may be under strong selection in response to local diseases. For example, people who are duffy negative tend to have higher resistance to malaria. Most Africans are duffy negative and most non-Africans are duffy positive. A number of genetic diseases more prevalent in malaria-afflicted areas may provide some genetic resistance to malaria including sickle cell disease, thalassaemias, glucose-6-phosphate dehydrogenase, and possibly others.

Many theories about the origin of the Cystic fibrosis have suggested that it provides a heterozygote advantage by giving resistance to diseases earlier common in Europe.

In earlier research a common theory was the "common disease-common variant" model. It argues that for common illnesses, the genetic contribution comes from the additive or multiplicative effects of gene variants that each one is common in the population. Each such gene variant is argued to cause only a small risk of disease and no single variant is enough to cause the disease. An individual must have many of these common gene variants in order for the risk of disease to be substantial.

More recent research indicates that the "common disease-rare variant" may be a better explanation for many common diseases. In this model, rare but higher-risk gene variants cause common diseases. This model may be particularly relevant for diseases that reduces fertility.

In contrast, for common genes associated with common disease to persist they must either have little effect during the reproductive period of life (like Alzheimer's disease) or provide some advantage in the original environment (like genes causing autoimmune diseases also providing resistance against infections). In either case varying frequencies of genes variants in different populations may be an explanation for health disparities. Genetic variants associated with Alzheimer's disease, deep venous thrombosis, Crohn disease, and type 2 diabetes appear to adhere to "common disease-common variant" model.

Gene Flow

Gene flow and admixture can also have an effect on relationships between race and race-linked disorders. Multiple sclerosis, for example, is typically associated with people of European descent, but due to admixture African Americans have elevated levels of the disorder relative to Africans. Some diseases and physiological variables vary depending upon their admixture ratios. Examples include measures of insulin functioning and obesity.

Gene Interactions

The same gene variant, or group of gene variants, may produce different effects in different populations depending on differences in the gene variants, or groups of gene variants, they interact with. One example is the rate of progression to AIDS and death in HIV–infected patients.

In Caucasians and Hispanics, HHC haplotypes were associated with disease retardation, particularly a delayed progression to death. In contrast, for African Americans, possession of HHC haplotypes was associated with disease acceleration.

Controversy Regarding Race in Biomedicine

There is a controversy regarding race as a method for classifying humans. Some it as a purely social construct. Others it as a biological reality reflecting average genetic group differences.

New interest in human biological variation has resulted in a resurgence of the use of race in biomedicine.

The main impetus for this development is the possibility of improving the prevention and treatment of certain diseases by predicting hard-to-ascertain factors, such as genetically conditioned health factors, on the basis of more easily ascertained characteristics such as phenotype and racial self-identification.

Since medical judgment often involves decision-making under uncertain conditions, many doctors consider it useful to take race into account when treating disease because diseases and treatment responses tend to cluster by geographic ancestry. The discovery that more diseases than previously thought correlate with racial identification have further sparked the interest in using race as a proxy for bio-geographical ancestry and genetic buildup.

Race in medicine is used as an approximation for more specific genetic and environmental risk factors. Race is thus partly a surrogate for environmental factors such as differences in socioeconomic status that are known to affect health.

It is also an imperfect surrogate for ancestral geographic regions and differences in gene frequencies between different ancestral populations and thus differences in genes that can affect health. This can give an approximation of probability for disease or for preferred treatment, although the approximation is less than perfect.

Taking the example of sickle-cell disease, in an emergency room, knowing the geographic origin of a patient may help a doctor doing an initial diagnosis if a patient presents with symptoms compatible with this disease. This is unreliable evidence with the disease being present in many different groups as noted above with the trait also present in some Mediterranean European populations. Definitive diagnosis comes from examining the blood of the patient. In the US, screening for sickle cell anemia is done on all newborns regardless of race.

The continued use of racial categories has been criticized. Apart from the general controversy regarding race, some argue that the continued use of racial categories in health care and as risk factors could result in increased stereotyping and discrimination in society and health services. On the other hand, also some of those who are critical of race as a biological concept see race as socially meaningful group that is important to study epidemiologically in order to reduce disparities. David Williams (1994) argued, after an examination of articles in the journal Health Services Research during the 1966-90 period, that how race was determined and defined was seldom described. At a minimum, researchers should describe if race was assessed by self-report, proxy report, extraction from records, or direct observation. Race was also often used questionable, such as an indicator of socioeconomic status. Racial genetic explanations may be overemphasized, ignoring the interaction with and the role of the environment.

There is general agreement that a goal of health-related genetics should be to move past the weak surrogate relationships of racial health disparity and get to the root causes of health and disease. This includes research which strives to analyze human genetic variation in smaller groups than races across the world. One such method is called ethnogenetic layering. It works by focusing on geographically identified microethnic groups. For example, in the Mississippi Delta region ethnogenetic layering might include such microethnic groups as the Cajun (as a subset of European Americans), the Creole and Black groups [with African origins in Senegambia, Central Africa and Bight of Benin] (as a subset of African Americans), and Choctaw, Houmas, Chickasaw, Coushatta, Caddo, Atakapa, Karankawa and Chitimacha peoples (as subsets of Native American Indians).

Better still may be individual genetic assessment of relevant genes. However, until cheaper and more widely available methods of genetic analysis are commonplace, the consideration of race remains a worthwhile practice for many doctors and researchers. Even when such methods become commonly available, race will continue to be important when looking at groups instead of individuals such as in epidemiologic research.

Association Studies

One area in which population categories can be important considerations in genetics research is in controlling for confounding between population genetic substructure, environmental exposures, and health outcomes. Association studies can produce spurious results if cases and controls have differing allele frequencies for genes that are not related to the disease being studied, although the magnitude of its problem in genetic association studies is subject to debate. Various techniques detect and account for population substructure, but these methods can be difficult to apply in practice.

Population genetic substructure also can aid genetic association studies. For example, populations that represent recent mixtures of separated ancestral groups can exhibit longer-range linkage disequilibrium between susceptibility alleles and genetic markers than is the case for other populations. Genetic studies can use this disequilibrium to search for disease alleles with fewer markers than would be needed otherwise. Association studies also can take advantage of the contrasting experiences of racial or ethnic groups, including migrant groups, to search for interactions between particular alleles and environmental factors that might influence health.

Human Genome Projects

The Human Genome Diversity Project has collected genetic samples from 52 indigenous populations. This sampling has been controversial because some fear that racists might use the results, because the sampling might overemphasize genetic distinctions, because of informed consent issues, and patenting issues.

Race and Health in the United States

Research on Race and health in the United States shows many health disparities between the different racial/ethnic groups. The possible causes, such as genetics, socioeconomic factors, and racism, continue to be debated.

Southern Blot

A Southern blot is a method routinely used in molecular biology for detection of a specific DNA sequence in DNA samples. Southern blotting combines transfer of electrophoresis-separated DNA fragments to a filter membrane and subsequent fragment detection by probe hybridization. The method is named after its inventor, the British biologist Edwin Southern. Other blotting methods (i.e., Western blot, Northern blot, Eastern blot, Southwestern blot) that employ similar principles, but using RNA or protein, have later been named in reference to Edwin Southern's name. As the technique was eponymously named, Southern blot is capitalized as is conventional for proper nouns. The names for other blotting methods may follow this convention, by analogy. Method

1. Restriction endonucleases are used to cut high-molecular-weight DNA strands into smaller fragments.
2. The DNA fragments are then electrophoresed on an agarose gel to separate them by size.

3. If some of the DNA fragments are larger than 15 kb, then prior to blotting, the gel may be treated with an acid, such as dilute HCl, which depurinates the DNA fragments, breaking the DNA into smaller pieces, thus allowing more efficient transfer from the gel to membrane.
4. If alkaline transfer methods are used, the DNA gel is placed into an alkaline solution (typically containing sodium hydroxide) to denature the double-stranded DNA. The denaturation in an alkaline environment may improve binding of the negatively charged DNA to a positively charged membrane, separating it into single DNA strands for later hybridization to the probe, and destroys any residual RNA that may still be present in the DNA. The choice of alkaline over neutral transfer methods, however, is often empirical and may result in equivalent results.
5. A sheet of nitrocellulose (or, alternatively, nylon) membrane is placed on top of (or below, depending on the direction of the transfer) the gel. Pressure is applied evenly to the gel (either using suction, or by placing a stack of paper towels and a weight on top of the membrane and gel), to ensure good and even contact between gel and membrane. If transferring by suction 20X SSC buffer is used to ensure a seal and prevent drying of the gel. Buffer transfer by capillary action from a region of high water potential to a region of low water potential (usually filter paper and paper tissues) is then used to move the DNA from the gel on to the membrane; ion exchange interactions bind the DNA to the membrane due to the negative charge of the DNA and positive charge of the membrane.
6. The membrane is then baked in a vacuum or regular oven at 80 °C for 2 hours (standard conditions; nitrocellulose or nylon membrane) or exposed to ultraviolet radiation (nylon membrane) to permanently attach the transferred DNA to the membrane.
7. The membrane is then exposed to a hybridization probe—a single DNA fragment with a specific sequence whose presence in the target DNA is to be determined. The probe DNA is labelled so that it can be detected, usually by incorporating radioactivity or tagging the molecule with a fluorescent or chromogenic dye. In some cases, the hybridization probe may be made from RNA, rather than DNA. To ensure the specificity of the binding of the probe to the sample DNA, most common hybridization methods use salmon or herring sperm DNA for blocking of the membrane surface and target DNA, deionized formamide, and detergents such as SDS to reduce non-specific binding of the probe.
8. After hybridization, excess probe is washed from the membrane (typically using SSC buffer), and the pattern of hybridization is visualized on X-ray film by autoradiography in the case of a radioactive or fluorescent probe, or by development of color on the membrane if a chromogenic detection method is used.

Result

Hybridization of the probe to a specific DNA fragment on the filter membrane indicates that this fragment contains DNA sequence that is complementary to the probe. The transfer step of the DNA from the electrophoresis gel to a membrane permits easy binding of the labeled hybridization probe to the size-fractionated DNA. It also allows for the fixation of the target-probe hybrids, required for analysis by autoradiography or other detection methods. Southern blots performed with restriction enzyme-digested genomic DNA may be used to determine the number of sequences (e.g., gene copies) in a genome.

A probe that hybridizes only to a single DNA segment that has not been cut by the restriction enzyme will produce a single band on a Southern blot, whereas multiple bands will likely be observed when the probe hybridizes to several highly similar sequences (e.g., those that may be the result of sequence duplication).

Modification of the hybridization conditions (for example, increasing the hybridization temperature or decreasing salt concentration) may be used to increase specificity and decrease hybridization of the probe to sequences that are less than 100% similar.

Triple-stranded DNA

A triple-stranded DNA is a structure of DNA in which three oligonucleotides wind around each other and form a triple helix. In this structure, one strand binds to a B-form DNA double helix through Hoogsteen or reversed Hoogsteen hydrogen bonds.

For example, a nucleobase T binds to a Watson–Crick base-pairing of T-A by Hoogsteen hydrogen bonds between an AxT pair (x represents a Hoogsteen base pair). An N-3 protonated cytosine, represented as C+, can also form a base-triplet with a C-G pair through the Hoogsteen base-pairing of an GxC+. Thus, the triple-helical DNAs using these Hoogsteen pairings consist of two homopyrimidines and one homopurine, and the homopyrimidine third strand is parallel to the homopurine strand.

A homopurine third strand can also bind to a homopurine-homopyrimidine duplex using reversed Hoogsteen patterns. In this triplex, a nucleobase A binds to a T-A base pair and a G to a C-G pair. Since the nucleobases on the third strand have to be reversed, the homopurine third strand is antiparallel to the homopurine strand of the original duplex.

Triple-stranded DNA was a common hypothesis in the 1950s when scientists were struggling to discover DNA's true structural from. Watson and Crick (who later won the Nobel Prize for their double-helix model) originally considered a triple-helix model, as did Pauling and Corey, who published a proposal for their triple-helix model in the 1953 scientific journal Nature, as well as fellow scientist Fraser. However, Watson and Crick soon identified several problems with these models:

1. Negatively charged phosphates near the axis will repel each other, leaving the question as to how the three-chain structure would stay together.
2. In a triple-helix model (specifically Pauling and Corey's model), some of the van der Waals distances appear to be too small.

Fraser's model differed from Pauling and Corey's in that in his model the phosphates are on the outside and the bases are on the inside, linked together by hydrogen bonds.

However, Watson and Crick found Fraser's model to be too ill-defined to comment specifically on its inadequacies in their publication in "Nature" (1953): Molecular Structure of Nucleic Acids. Triple-stranded DNA was also described in 1957, when it was thought to occur in only one in vivo biological process: as an intermediate product during the action of the E. coli recombination enzyme RecA. Its role in that process is not understood.

Using nucleic acid segments that bind to the DNA duplexes to form triple strands as a way of regulating gene expression is under investigation by biotechnology companies. Similar work is also being undertaken at Yale University.

Triple-stranded DNA has been implicated in the regulation of several genes. One, the c-myc gene has been extensively mutated to examine the exact role that triplex DNA structure, versus linear sequence, plays in gene regulation. A c-myc promoter element, termed the nuclease-sensitive element or NSE, can form tandem intramolecular triplexes of the H-DNA type and has a repeating sequence motif (ACCCTCCCC)4.

The NSE when mutated and examined for transcriptional activity and for intra- and intermolecular triplex forming ability. The transcriptional activity of mutant NSEs can be predicted by the element's ability to form H-DNA and not by repeat number, position, or the number of mutant base pairs. DNA may therefore be a dynamic participant in the transcription of the c-myc gene.

X-ray Scattering Techniques

X-ray scattering techniques are a family of non-destructive analytical techniques which reveal information about the crystallographic structure, chemical composition, and physical properties of materials and thin films.

These techniques are based on observing the scattered intensity of an X-ray beam hitting a sample as a function of incident and scattered angle, polarization, and wavelength or energy.

X-ray Diffraction Techniques

X-ray diffraction yields the atomic structure of materials and is based on the elastic scattering of X-rays from the electron clouds of the individual atoms in the system. The most comprehensive description of scattering from crystals is given by the dynamical theory of diffraction.

- Single-crystal X-ray diffraction is a technique used to solve the complete structure of crystalline materials, ranging from simple inorganic solids to complex macromolecules, such as proteins.
- Powder diffraction (XRD) is a technique used to characterise the crystallographic structure, crystallite size (grain size), and preferred orientation in polycrystalline or powdered solid samples. Powder diffraction is commonly used to identify unknown substances, by comparing diffraction data against a database maintained by the International Centre for Diffraction Data. It may also be used to characterize heterogeneous solid mixtures to determine relative abundance of crystalline compounds and, when coupled with lattice refinement techniques, such as Rietveld refinement, can provide structural information on unknown materials. Powder diffraction is also a common method for determining strains in crystalline materials. An effect of the finite crystallite sizes is seen as a broadening of the peaks in an X-ray diffraction as is explained by the Scherrer Equation.
- Thin film diffraction and grazing incidence X-ray diffraction may be used to characterize the crystallographic structure and preferred orientation of substrate-anchored thin films.
- High-resolution X-ray diffraction is used to characterize thickness, crystallographic structure, and strain in thin epitaxial films. It employs parallel-beam optics.
- X-ray pole figure analysis enables one to analyze and determine the distribution of crystalline orientations within a crystalline thin-film sample.
- X-ray rocking curve analysis is used to quantify grain size and mosaic spread in crystalline materials.

6

Prognosis and Treatment of Genetic Disorders

Genetic disorders rarely have effective treatments, though gene therapy is being tested as a possible treatment for some genetic diseases, including some forms of retinitis pigmentosa Huntington's disease clinical research

There are different therapies under investigation for Huntington's disease. Huntington's disease is a genetic neurological disorder characterized after onset by uncoordinated, jerky body movements and a decline in some mental abilities for which there is no cure or effective treatments. Animal models help in the ongoing investigations using intrabody therapy, gene silencing, or stem cell implants, but all are at their first steps. Several clinical trials of various compounds are also in development. Many of these trials use the unified Huntington's disease rating scale (UHDRS) developed by the Huntington's Study Group. This provides an overall rating system based on motor, behavioral, cognitive, and functional assessments, backed up by a reference database of previous results provided by HSG for reference. Another avenue of research is understanding the physical mechanics of protein folding, HD is one of several targets of the folding@home project, one of the largest distributed computing systems on the World Wide Web. Animal models

Appropriate animal models are critical for understanding the fundamental mechanisms causing the disease and for supporting the early stages of drug development. Neurochemically-induced mice or monkeys were first available, but they did not mimic the progressive features of the disease. After the HD gene was discovered, transgenic animals exhibiting HD were generated by inserting a CAG repeat expansion into the genome, of mice (strain R6/2),), Drosophila fruit flies, and more recently monkeys. Expression without insertion of a DNA repeat in nematode worms also produced a valuable model.

Intrabody Therapy

Genetically-engineered intracellular antibody fragments called intrabodies have shown therapeutic results, by inhibiting mHtt aggregation, in Drosophila models. This was achieved using an intrabody called C4 sFv, a single chain variable fragment which binds to the end of mHtt within a cell. This therapy prevented larval and pupal mortality (without therapy 77% died) and delayed neurodegeneration in the adult, significantly increasing their lifespan. Intrabodies have shown promise as a useful tool for drug discovery, and may potentially be used as a therapy for HD and other neurodegenerative disorders caused by protein mis-folding or abnormal protein interactions.

Gene Silencing

As HD has been conclusively linked to a single gene, gene silencing is potentially possible. Researchers have investigated using gene knockdown of mHTT as a potential treatment. Using a mouse model, siRNA therapy achieved a 60% reduction in expression of mHTT and progression of the disease was stalled. However, this study used the human form of the mHTT protein in mice, and was therefore only able to directly target the mHTT, leaving endogenous, wild-type mouse Htt gene expression unaffected. The widely-held hypothesis that silencing the wild type mouse Htt in middle aged mice, the age of onset of HD, renders undesirable, gene silencing to treat HD, is now largely discredited by experiment.

These data indicate that the coincident silencing of the wild-type and mutant htt may be considered as a therapeutic tool for HD. This is despite the unchallenged experimental evidence that wild-type Htt is critical for development in mouse models. Another study showed that mouse models already in late stages of the disease recovered motor function after expression of mHTT was stopped.

Stem Cell Treatments

Stem cell treatments are a type of intervention strategy that introduces new cells into damaged tissue in order to treat disease or injury. Many medical researchers believe that stem cell treatments have the potential to change the face of human disease and alleviate suffering. The ability of stem cells to self-renew and give rise to subsequent generations with variable degrees of differentiation capacities, offers significant potential for generation of tissues that can potentially replace diseased and damaged areas in the body, with minimal risk of rejection and side effects.

A number of stem cell therapies exist, but most are at experimental stages or costly, with the notable exception of bone marrow transplantation. Medical researchers anticipate that adult and embryonic stem cells will soon be able to treat cancer, Type 1 diabetes mellitus, Parkinson's disease, Huntington's disease, Celiac Disease, cardiac failure, muscle damage and neurological disorders, and many others. Nevertheless, before stem cell therapeutics can be applied in the clinical setting, more research is necessary to understand stem cell behavior upon transplantation as well as the mechanisms of stem cell interaction with the diseased/injured microenvironment.

Stem cell implants are a stem cell treatment where damaged neurons are replaced by transplantation of stem cells (or possibly neural stem cells—a type of somatic [adult] stem cell) into affected regions of the brain. Hypothetically, embryonic stem cells can be differentiated into neuronal precursors in vitro, and transplanted into damaged areas of the brain to generate replacement neurons; if enough damaged neurons can be replaced and develop the correct synaptic connectivity, symptoms could be alleviated. This treatment would not prevent further neuronal damage, so it would have to be an ongoing treatment. Experiments have yielded some positive results in animal models, but stem cell treatment for HD remains highly speculative and has not been tested in clinical trials. Current treatments

For over 30 years, bone marrow, and more recently, umbilical cord blood stem cells, have been used to treat cancer patients with conditions such as leukemia and lymphoma. During chemotherapy, most growing cells are killed by the cytotoxic agents. These agents, however, cannot discriminate between the leukemia or neoplastic cells, and the hematopoietic stem cells within the bone marrow. It is this side effect of conventional chemotherapy strategies that the stem cell transplant attempts to reverse; a donor's healthy bone marrow reintroduces functional stem cells to replace the cells lost in the host's body during treatment.

Hematopoietic Stem Cell Transplantation

Hematopoietic stem cell transplantation (HSCT) is the transplantation of multipotent hematopoietic stem cell or blood, usually derived from bone marrow, peripheral blood stem cells, or umbilical cord blood. Stem cell transplantation is a medical procedure in the fields of hematology and oncology, most often performed for people with diseases of the blood, bone marrow, or certain cancers.

Hematopoietic stem cell transplantation remains a risky procedure with many possible complications; it has traditionally been reserved for patients with life-threatening diseases. While occasionally used experimentally in nonmalignant and nonhematologic indications such as severe disabling autoimmune disease and cardiovascular disease, the risk of fatal complications appears too high to gain wider acceptance. A total of 50,417 first hematopoietic stem cell transplants were reported as taking place worldwide in 2006, according to a global survey of 1327 centers in 71 countries conducted by the Worldwide Network for Blood and Marrow Transplantation. Of these, 28,901 (57%) were autologous and 21,516 (43%) were allogeneic (11,928 from family donors and 9,588 from unrelated donors).

The main indications for transplant were lymphoproliferative disorders (54.5%) and leukemias (33.8%), and the majority took place in either Europe (48%) or the Americas (36%). In 2009, according to the world marrow donor association, stem cell products provided for unrelated transplantation worldwide had increased to 15,399 (3,445 bone marrow donations, 8,162 peripheral blood stem cell donations, and 3,792 cord blood units).

While it remains a procedure involving a high degree of risk, one must keep in mind that many disorders (mainly acute leukemias and neoplastic lymphoproliferative disorders) treated by this procedure—which is usually not attempted unless necessary because of prior treatment failures or the patient's emergent condition (unresponsive or severe acute leukemia, relapsed acute leukemia, or progressing, formerly chronic leukemias that have not responded to other regimens and have entered blast crisis)- themselves will usually carry a mortality and morbidity rate that is as high and often higher than that for the procedure.

Further, less severe complications from this procedure, and even those that carry a worse prognosis such as many cases of acute graft versus host disease, can be dealt with at least to a point, if the patient was in acceptable health beforehand and the condition was noticed and aggressively treated early, with modern standards of tertiary medical care.

History

Georges Mathé, a French oncologist, performed the first bone marrow transplant in 1959 on five Yugoslavian nuclear workers whose own marrow had been damaged by irradiation caused by a Criticality accident at the Vinèa Nuclear Institute, but all of these transplants were rejected. Mathé later pioneered the use of bone marrow transplants in the treatment of leukemia.

Stem cell transplantation was pioneered using bone-marrow-derived stem cells by a team at the Fred Hutchinson Cancer Research Center from the 1950s through the 1970s led by E. Donnall Thomas, whose work was later recognized with a Nobel Prize in Physiology or Medicine. Thomas' work showed that bone marrow cells infused intravenously could repopulate the bone marrow and produce new blood cells. His work also reduced the likelihood of developing a life-threatening complication called graft-versus-host disease.

The first physician to perform a successful human bone marrow transplant on a disease other than cancer was Robert A. Good at the University of Minnesota in 1968.

Indications

Many recipients of HSCTs are multiple myeloma or leukemia patients who would not benefit from prolonged treatment with, or are already resistant to, chemotherapy. Candidates for HSCTs include pediatric cases where the patient has an inborn defect such as severe combined immunodeficiency or congenital neutropenia with defective stem cells, and also children or adults with aplastic anemia who have lost their stem cells after birth.

Other conditions treated with stem cell transplants include sickle-cell disease, myelodysplastic syndrome, neuroblastoma, lymphoma, Ewing's Sarcoma, Desmoplastic small round cell tumor, chronic granulomatous disease and Hodgkin's disease.

More recently non-myeloablative, or so-called "mini transplant," procedures have been developed that require smaller doses of preparative chemo and radiation. This has allowed HSCT to be conducted in the elderly and other patients who would otherwise be considered too weak to withstand a conventional treatment regimen.

List of conditions treated with hematopoietic stem cell transplantation

Hematopoietic stem cell transplantation may be used to treat a number of conditions both congenital and acquired.

Acquired

- Malignancies
    - o Hematological
        - § Leukemias
        - § Acute lymphoblastic leukemia (ALL)
        - § Acute myeloid leukemia (AML)
        - § Chronic lymphocytic leukemia (CLL)
        - § Chronic myelogenous leukemia (CML), accelerated phase or blast crisis
        - § Lymphomas
        - § Hodgkin's disease
        - § Non-Hodgkin's lymphoma
        - § Myelomas
        - § Multiple myeloma (Kahler's disease)

- o Solid tumor cancers
  - § Neuroblastoma
  - § Desmoplastic small round cell tumor
  - § Ewing's sarcoma
  - § Choriocarcinoma
- Hematologic disease
  - o Phagocyte disorders
    - § Myelodysplasia
  - o Anemias
    - § Paroxysmal nocturnal hemoglobinuria (PNH; severe aplasia)
    - § Aplastic anemia
    - § Acquired pure red cell aplasia
  - o Myeloproliferative disorders
    - § Polycythemia vera
    - § Essential thrombocytosis
    - § Myelofibrosis
- Metabolic disorders
  - o Amyloidoses
    - § Amyloid light chain (AL) amyloidosis
- Environmentally-induced diseases
  - o Radiation poisoning
- Viral diseases
  - o HIV

Congenital

- Lysosomal storage disorders
  - o Lipidoses (disorders of lipid storage)
    - § Neuronal ceroid lipofuscinoses
    - § Infantile neuronal ceroid lipofuscinosis (INCL, Santavuori disease,)
    - § Jansky-Bielschowsky disease (late infantile neuronal ceroid lipofuscinosis)
    - § Sphingolipidoses
    - § Niemann-Pick disease

§ Gaucher disease

§ Leukodystrophies

§ Adrenoleukodystrophy

§ Metachromatic leukodystrophy

§ Krabbe disease (globoid cell leukodystrophy)

o Mucopolysaccharidoses

§ Hurler syndrome (MPS I H, á-L-iduronidase deficiency)

§ Scheie syndrome (MPS I S)

§ Hurler-Scheie syndrome (MPS I H-S)

§ Hunter syndrome (MPS II, iduronidase sulfate deficiency)

§ Sanfilippo syndrome (MPS III)

§ Morquio syndrome (MPS IV)

§ Maroteaux-Lamy syndrome (MPS VI)

§ Sly syndrome (MPS VII)

o Glycoproteinoses

§ Mucolipidosis II (I-cell disease)

§ Fucosidosis

§ Aspartylglucosaminuria

§ Alpha-mannosidosis

o Other

§ Wolman disease (acid lipase deficiency)

• Immunodeficiencies

o T-cell deficiencies

§ Ataxia telangiectasia

§ DiGeorge syndrome

o Combined T- and B-cell deficiencies

§ Severe combined immunodeficiency (SCID), all types

o Well-defined syndromes

§ Wiskott-Aldrich syndrome

o Phagocyte disorders

§ Kostmann syndrome

§ Shwachman-Diamond syndrome

- o Immune dysregulation diseases
  - § Griscelli syndrome, type II
- o Innate immune deficiencies
  - § NF-Kappa-B Essential Modulator (NEMO) deficiency (Inhibitor of Kappa Light Polypeptide Gene Enhancer in B Cells Gamma Kinase deficiency)
- Hematologic diseases
  - § Hemoglobinopathies
  - § Sickle cell disease
  - § â thalassemia major (Cooley's anemia)
- o Anemias
  - § Aplastic anemia
  - § Diamond-Blackfan anemia
  - § Fanconi anemia
- o Cytopenias
  - § Amegakaryocytic thrombocytopenia
- o Hemophagocytic syndromes
  - § Hemophagocytic lymphohistiocytosis (HLH)

Graft Types

Autologous

Autologous HSCT requires the extraction (apheresis) of haematopoietic stem cells (HSC) from the patient and storage of the harvested cells in a freezer. The patient is then treated with high-dose chemotherapy with or without radiotherapy with the intention of eradicating the patient's malignant cell population at the cost of partial or complete bone marrow ablation (destruction of patient's bone marrow function to grow new blood cells).

The patient's own stored stem cells are then returned to his/her body, where they replace destroyed tissue and resume the patient's normal blood cell production. Autologous transplants have the advantage of lower risk of infection during the immune-compromised portion of the treatment since the recovery of immune function is rapid. Also, the incidence of patients experiencing rejection (graft-versus-host disease) is very rare due to the donor and recipient being the same individual. These advantages have established autologous HSCT as one of the standard second-line treatments for such diseases as lymphoma. However, for others such as Acute Myeloid Leukemia, the reduced mortality of the autogenous relative to allogeneic HSCT may be outweighed by an increased likelihood of cancer relapse and related mortality, and therefore the allogeneic treatment may be preferred for those conditions. Researchers have conducted small studies using non-myeloablative hematopoietic stem cell transplantation as a possible treatment for type I (insulin dependent) diabetes in children and adults. Results have been promising; however, as of 2009 it was premature to speculate whether these experiments will lead to effective treatments for diabetes.

Allogeneic

Allogeneic HSCT involves two people: the (healthy) donor and the (patient) recipient. Allogeneic HSC donors must have a tissue (HLA) type that matches the recipient. Matching is performed on the basis of variability at three or more loci of the HLA gene, and a perfect match at these loci is preferred. Even if there is a good match at these critical alleles, the recipient will require immunosuppressive medications to mitigate graft-versus-host disease. Allogeneic transplant donors may be related (usually a closely HLA matched sibling), syngeneic (a monozygotic or 'identical' twin of the patient - necessarily extremely rare since few patients have an identical twin, but offering a source of perfectly HLA matched stem cells) or unrelated (donor who is not related and found to have very close degree of HLA matching). Unrelated donors may be found through a registry of bone marrow donors such as the National Marrow Donor Program. People who would like to be tested for a specific family member or friend without joining any of the bone marrow registry data banks may contact a private HLA testing laboratory and be tested with a mouth swab to see if they are a potential match.

A "savior sibling" may be intentionally selected by preimplantation genetic diagnosis in order to match a child both regarding HLA type and being free of any obvious inheritable disorder. Allogeneic transplants are also performed using umbilical cord blood as the source of stem cells. In general, by transplanting healthy stem cells to the recipient's immune system, allogeneic HSCTs appear to improve chances for cure or long-term remission once the immediate transplant-related complications are resolved.

A compatible donor is found by doing additional HLA-testing from the blood of potential donors. The HLA genes fall in two categories (Type I and Type II). In general, mismatches of the Type-I genes (i.e. HLA-A, HLA-B, or HLA-C) increase the risk of graft rejection. A mismatch of an HLA Type II gene (i.e. HLA-DR, or HLA-DQB1) increases the risk of graft-versus-host disease. In addition a genetic mismatch as small as a single DNA base pair is significant so perfect matches require knowledge of the exact DNA sequence of these genes for both donor and recipient. Leading transplant centers currently perform testing for all five of these HLA genes before declaring that a donor and recipient are HLA-identical. Race and ethnicity are known to play a major role in donor recruitment drives, as members of the same ethnic group are more likely to have matching genes, including the genes for HLA.

Sources and Storage of Cells

To limit the risks of transplanted stem cell rejection or of severe graft-versus-host disease in allogeneic HSCT, the donor should preferably have the same human leukocyte antigens (HLA) as the recipient. About 25 to 30 percent of allogeneic HSCT recipients have an HLA-identical sibling. Even so-called "perfect matches" may have mismatched minor alleles that contribute to graft-versus-host disease.

Bone Marrow

In the case of a bone marrow transplant, the HSC are removed from a large bone of the donor, typically the pelvis, through a large needle that reaches the center of the bone. The technique is referred to as a bone marrow harvest and is performed under general anesthesia.

Peripheral Blood Stem Cells

Peripheral stem cell transplantation is a method of replacing blood-forming cells destroyed by cancer treatment. Immature blood cells (stem cells) in the circulating blood that are similar to those in the bone marrow are given to the patient after treatment. This helps the bone marrow recover and continue producing healthy blood cells. Transplantation may be autologous (an individual's own blood cells saved earlier), allogeneic (blood cells donated by someone else), or syngeneic (blood cells donated by an identical twin). Also called peripheral stem cell support.

Peripheral blood stem cells are now the most common source of stem cells for allogeneic HSCT. They are collected from the blood through a process known as apheresis. The donor's blood is withdrawn through a sterile needle in one arm and passed through a machine that removes white blood cells. The red blood cells are returned to the donor. The peripheral stem cell yield is boosted with daily subcutaneous injections of Granulocyte-colony stimulating factor, serving to mobilize stem cells from the donor's bone marrow into the peripheral circulation.

Amniotic Fluid

It is also possible to extract hematopoietic stem cells from amniotic fluid for both autologous or heterologous use at the time of childbirth.

Umbilical Cord Blood

Umbilical cord blood is obtained when a mother donates her infant's umbilical cord and placenta after birth. Cord blood has a higher concentration of HSC than is normally found in adult blood. However, the small quantity of blood obtained from an umbilical cord (typically about 50 mL) makes it more suitable for transplantation into small children than into adults. Newer techniques using ex-vivo expansion of cord blood units or the use of two cord blood units from different donors are being explored to allow cord blood transplants to be used in adults.

Cord blood is typically harvested from the umbilical cord of a child being born after preimplantation genetic diagnosis (PGD) for human leucocyte antigen (HLA) matching in order to donate to an ill sibling requiring HSCT.

Storage of HSC

Unlike other organs, bone marrow cells can be frozen (cryopreserved) for prolonged periods, without damaging too many cells. This is a necessity with autologous HSC because the cells must be harvested from the recipient months in advance of the transplant treatment. In the case of allogeneic transplants, fresh HSC are preferred, in order to avoid cell loss that might occur during the freezing and thawing process. Allogeneic cord blood is stored frozen at a cord blood bank because it is only obtainable at the time of childbirth. To cryopreserve HSC, a preservative, DMSO, must be added, and the cells must be cooled very slowly in a controlled-rate freezer to prevent osmotic cellular injury during ice crystal formation. HSC may be stored for years in a cryofreezer, which typically utilizes liquid nitrogen.

Conditioning Regimens

Myeloablative Transplants

The chemotherapy or irradiation given immediately prior to a transplant is called the conditioning or preparative regimen, the purpose of which is to help eradicate the patient's disease prior to the infusion of HSC and to suppress immune reactions. The bone marrow can be ablated with dose-levels that cause minimal injury to other tissues. In allogeneic transplants a combination of cyclophosphamide with busulfan or total body irradiation is commonly employed. This treatment also has an immunosuppressive effect that prevents rejection of the HSC by the recipient's immune system. The post-transplant prognosis often includes acute and chronic graft-versus-host disease that may be life-threatening. However in certain leukemias this can coincide with protection against cancer relapse owing to the graft versus tumor effect. Autologous transplants may also use similar conditioning regimens, but many other chemotherapy combinations can be used depending on the type of disease.

Non-myeloablative Allogeneic Transplants

This is a newer treatment approach using lower doses of chemotherapy and radiation, which are too low to eradicate all the bone marrow cells of a recipient. Instead, non-myeloablative transplants run lower risks of serious infections and transplant-related mortality while relying upon the graft versus tumor effect to resist the inherent increased risk of cancer relapse.

Also significantly, while requiring high doses of immunosuppressive agents in the early stages of treatment, these doses are less than for conventional transplants. This leads to a state of mixed chimerism early after transplant where both recipient and donor HSC coexist in the bone marrow space.

Decreasing doses of immunosuppressive therapy then allows donor T-cells to eradicate the remaining recipient HSC and to induce the graft versus tumor effect. This effect is often accompanied by mild graft-versus-host disease, the appearance of which is often a surrogate for the emergence of the desirable graft versus tumor effect, and also serves as a signal to establish an appropriate dosage level for sustained treatment with low levels of immunosuppressive agents.

Because of their gentler conditioning regimens, these transplants are associated with a lower risk of transplant-related mortality and therefore allow patients who are considered too high-risk for conventional allogeneic HSCT to undergo potentially curative therapy for their disease. These new transplant strategies are still somewhat experimental, but are being used more widely on elderly patients unfit for myeloablative regimens and for whom the higher risk of cancer relapse may be acceptable.

Engraftment

After several weeks of growth in the bone marrow, expansion of HSC and their progeny is sufficient to normalize the blood cell counts and reinitiate the immune system. The offspring of donor-derived hematopoietic stem cells have been documented to populate many different organs of the recipient, including the heart, liver, and muscle, and these cells had been suggested to have the abilities of regenerating injured tissue in these organs. However, recent research has shown that such lineage infidelities does not occur as a normal phenomenon.

Complications

HSCT is associated with a high treatment-related mortality in the recipient (10% or higher), which limits its use to conditions that are themselves life-threatening. Major complications are veno-occlusive disease, mucositis, infections (sepsis), graft-versus-host disease and the development of new malignancies.

Infection

Bone marrow transplantation usually requires that the recipient's own bone marrow be destroyed ("myeloablation"). Prior to "engraftment" patients may go for several weeks without appreciable numbers of white blood cells to help fight infection. This puts a patient at high risk of infections, sepsis and septic shock, despite prophylactic antibiotics. However, antiviral medications, such as acyclovir and valacyclovir, are quite effective in prevention of HSCT-related outbreak of herpetic infection in seropositive patients. The immunosuppressive agents employed in allogeneic transplants for the prevention or treatment of graft-versus-host disease further increase the risk of opportunistic infection. Immunosuppressive drugs are given for a minimum of 6-months after a transplantation, or much longer if required for the treatment of graft-versus-host disease. Transplant patients lose their acquired immunity, for example immunity to childhood diseases such as measles or polio. For this reason transplant patients must be re-vaccinated with childhood vaccines once they are off immunosuppressive medications.

Veno-occlusive Disease

Severe liver injury can result from hepatic veno-occlusive disease (VOD). Elevated levels of bilirubin, hepatomegaly and fluid retention are clinical hallmarks of this condition. There is now a greater appreciation of the generalized cellular injury and obstruction in hepatic vein sinuses, and hepatic VOD has lately been referred to as sinusoidal obstruction syndrome (SOS). Severe casesof SOS are associated with a high mortality rate. Anticoagulants or defibrotide may be effective in reducing the severity of VOD but may also increase bleeding complications. Ursodiol has been shown to help prevent VOD, presumably by facilitating the flow of bile.

Mucositis

The injury of the mucosal lining of the mouth and throat is a common regimen-related toxicity following ablative HSCT regimens. It is usually not life-threatening but is very painful, and prevents eating and drinking. Mucositis is treated with pain medications plus intravenous infusions to prevent dehydration and malnutrition.

Graft-versus-host Disease

Graft-versus-host disease (GVHD) is a common complication after a stem cell transplant or bone marrow transplant from another person (an allogeneic transplant). Immune cells (white blood cells) in the donated marrow or stem cells (the graft) recognize the recipient (the host) as "foreign". The transplanted immune cells then attack the host's body cells. Graft versus host disease can also take place during a blood transfusion under certain circumstances.

Graft-versus-host disease (GVHD) is an inflammatory disease that is unique to allogeneic transplantation. It is an attack of the "new" bone marrow's immune cells against the recipient's tissues. This can occur even if the donor and recipient are HLA-identical because the immune system can still recognize other differences between their tissues. It is aptly named graft-versus-host disease because bone marrow transplantation is the only transplant procedure in which the transplanted cells must accept the body rather than the body accepting the new cells. Acute graft-versus-host disease typically occurs in the first 3 months after transplantation and may involve the skin, intestine, or the liver.. High-dose corticosteroids such as prednisone are a standard treatment; however this immuno-suppressive treatment often leads to deadly infections. Chronic graft-versus-host disease may also develop after allogeneic transplant. It is the major source of late treatment-related complications, although it less often results in death. In addition to inflammation, chronic graft-versus-host disease may lead to the development of fibrosis, or scar tissue, similar to scleroderma; it may cause functional disability and require prolonged immunosuppressive therapy. Graft-versus-host disease is usually mediated by T cells, which react to foreign peptides presented on the MHC of the host·

Causes

According to the 1966 Billingham Criteria, 3 criteria must be met in order for GVHD to occur.

- An immuno-competent graft is administered, with viable and functional immune cells.
- The recipient is immunologically disparate - histo-incompatible.
- The recipient is immuno-compromised and therefore cannot destroy or inactivate the transplanted cells.

After bone marrow transplantation, T cells present in the graft, either as contaminants or intentionally introduced into the host, attack the tissues of the transplant recipient after perceiving host tissues as antigenically foreign.

The T cells produce an excess of cytokines, including TNF-á and interferon-gamma (IFNã). A wide range of host antigens can initiate graft-versus-host-disease, among them the human leukocyte antigens (HLAs). However, graft-versus-host disease can occur even when HLA-identical siblings are the donors. HLA-identical siblings or HLA-identical unrelated donors often have genetically different proteins (called minor histocompatibility antigens) that can be presented by Major histocompatibility complex (MHC) molecules to the donor's T-cells, which see these antigens as foreign and so mount an immune response.

While donor T-cells are undesirable as effector cells of graft versus host disease, they are valuable for engraftment by preventing the recipient's residual immune system from rejecting the bone marrow graft (host-versus-graft). In addition, as bone marrow transplantation is frequently used to treat cancer, mainly leukemias, donor T-cells have proven to have a valuable graft-versus-tumor effect. A great deal of current research on allogeneic bone marrow transplantation involves attempts to separate the undesirable graft-vs-host-disease aspects of T-cell physiology from the desirable graft-versus-tumor effect.

Types

In the clinical setting, graft-versus-host-disease is divided into acute and chronic forms.

- The acute or fulminant form of the disease (aGVHD) is normally observed within the first 100 days post-transplant, and is a major challenge to transplants owing to associated morbidity and mortality.
- The chronic form of graft-versus-host-disease (cGVHD) normally occurs after 100 days. The appearance of moderate to severe cases of cGVHD adversely influences long-term survival.

This distinction is not arbitrary: Acute and chronic graft-versus-host-disease appear to involve different immune cell subsets, different cytokine profiles, somewhat different host targets, and respond differently to treatment.

Clinical Manifestation

In the classical sense, acute graft-versus-host-disease is characterized by selective damage to the liver, skin (rash), and mucosa, and the gastrointestinal tract. Newer research indicates that other graft-versus-host-disease target organs include the immune system (the hematopoietic system, e.g., the bone marrow and the thymus) itself, and the lungs in the form of idiopathic pneumonitis. Chronic graft-versus-host-disease also attacks the above organs, but over its long-term course can also cause damage to the connective tissue and exocrine glands.

Acute GVHD of the GI tract can result in severe intestinal inflammation, sloughing of the mucosal membrane, severe diarrhea, abdominal pain, nausea, and vomiting. This is typically diagnosed via intestinal biopsy. Liver GVHD is measured by the bilirubin level in acute patients. Skin GVHD results in a diffuse maculopapular rash, sometimes in a lacy pattern.

It appears in both acute and chronic GVHD, mucosal damage to the vagina can result in severe pain and scarring. This can result in an inability to have sexual intercourse. Acute GVHD is staged as follows: overall grade (skin-liver-gut) with each organ staged individually from a low of 1 to a high of 4. Patients with grade IV GVHD usually have a poor prognosis.

If the GVHD is severe and requires intense immunosuppression involving steroids and additional agents to get under control, the patient may develop severe infections as a result of the immunosuppression and may die of infection.

In the oral cavity, chronic graft-versus-host-disease manifests as lichen planus with a higher risk of malignant transformation to oral squamous cell carcinoma in comparison to the classical oral lichen planus. Graft-versus-host-disease-associated oral cancer may have more aggressive behavior with poorer prognosis, when compared to oral cancer in non-hematopoietic stem cell transplantation patients.

Transfusion-associated GVHD

Transfusion-associated graft versus host disease (TA-GvHD) is a rare complication of blood transfusion, in which the donor T lymphocytes mount an immune response against the recipient's lymphoid tissue. Donor lymphocytes are usually identified as foreign and destroyed by the recipient's immune system.

However, in situations where the recipient is immunocompromised (inborn immunodeficiency, acquired immunodeficiency, malignancy), or when the donor is homozygous and the recipient is heterozygous for an HLA haplotype (as can occur in directed donations from first-degree relatives), the recipient's immune system is not able to destroy the donor lymphocytes. This can result in graft versus host disease.

This type of GVHD is associated with transfusion of un-irradiated blood to immunocompromised recipients. It can also occur in situations in which the blood donor is homozygous and the recipient is heterozygous for an HLA haplotype. It is associated with higher mortality (80-90%) due to involvement of bone marrow lymphoid tissue, however the clinical manifestations are similar to GVHD resulting from bone marrow transplantation. Transfusion-associated GVHD is rare in modern medicine. It is almost entirely preventable by controlled irradiation of blood products to inactivate the white blood cells (including lymphocytes) within. Epidemiology and pathogenesis

The incidence in immunocompromised patients receiving blood transfusions is estimated to be 0.1 - 1.0%, mortality around 80 - 90%. Mortality is higher in TA-GvHD than in GvHD associated with bone marrow transplantation, where the engrafted lymphoid cells in the bone marrow are of donor origin; therefore, the immune reaction is not directed against them. The most common cause of death in TA-GvHD is infections and hemorrhages, secondary to pancytopenia and liver dysfunction.

Presentation and Diagnosis

Clinical Manifestations

The clinical presentation is the same as GvHD occurring in other settings, such as bone marrow transplantation. TA-GvHD can develop four to thirty days after the transfusion. Typical symptoms include:

- fever
- erythematous maculopapular rash, which can progress to generalised erythroderma
- toxic epidermal necrolysis in extreme cases

Other symptoms can include cough, abdominal pain, vomiting, and profuse diarrhea (up to 8 liters/day).

Laboratory Manifestations

Laboratory findings include pancytopenia, abnormal liver enzymes, and electrolyte imbalance (when diarrhea is present).

Diagnosis

TA-GvHD can be suspected from a biopsy of the affected skin, and established by HLA analysis of the circulating lymphocytes. This testing can identify circulating lymphocytes with a different HLA type than the tissue cells of the host.

Treatment and Prevention

Treatment is only supportive, as no available form of therapy has proven effective in treating TA-GvHD.

Prevention includes gamma irradiation of the lymphocyte-containing blood products. This procedure should be performed in transfusions when:

- the recipient is immunocompromised
- the blood components are from a family donor
- HLA-matched platelets are transferred.

Another means of prevention is the use of third- or fourth-generation leukoreduction filters, although the efficacy of this procedure has not yet been documented.

In Thymus Transplantation

Thymus transplantation may be said to be able to cause a special type of GVHD because the recipients thymocytes would use the donor thymus cells as models when going through the negative selection to recognize self-antigens, and could therefore still mistake own structures in the rest of the body for being non-self.

This is a rather indirect GVHD because it is not directly cells in the graft itself that causes it but cells in the graft that make the recipient's T cells act like donor T cells. It can be seen as a multiple-organ autoimmunity in xenotransplantation experiments of the thymus between different species. Autoimmune disease is a frequent complication after human allogeneic thymus transplantation, found in 42% of subjects over 1 year post transplantation. However, this is partially explained by the fact that the indication itself, that is, complete DiGeorge syndrome, increases the risk of autoimmune disease.

Prevention

- DNA-based tissue typing allows for more precise HLA matching between donors and transplant patients, which has been proven to reduce the incidence and severity of GVHD and to increase long-term survival.
- The T-cells of umbilical cord blood (UCB) have an inherent immunological immaturity, and the use of UCB stem cells in unrelated donor transplants has a reduced incidence and severity of GVHD.
- Methotrexate, ciclosporin and tacrolimus are common drugs used for GVHD prophylaxis.
- Graft-versus-host-disease can largely be avoided by performing a T-cell-depleted bone marrow transplant. However, these types of transplants come at a cost of diminished graft-versus-tumor effect, greater risk of engraftment failure, or cancer relapse, and general immunodeficiency, resulting in a patient more susceptible to viral, bacterial, and fungal infection. In a multi-center study, disease-free survival at 3 years was not different between T cell-depleted and T cell-replete transplants.

Treatment of GVHD

Intravenously administered corticosteroids, such as prednisone, are the standard of care in acute GVHD and chronic GVHD. The use of these corticosteroids is designed to suppress the T-cell-mediated immune onslaught on the host tissues; however, in high doses, this immune-suppression raises the risk of infections and cancer relapse.

Therefore, it is desirable to taper off the post-transplant high-level steroid doses to lower levels, at which point the appearance of mild GVHD may be welcome, especially in HLA mis-matched patients, as it is typically associated with a graft-versus-tumor effect.

The use of irradiated blood components prevent GVHD. The irradiation of blood eliminates the proliferative capacity of lymphocytes present in red cell, platelet and freshly collected plasma components. The penetrated photons of radiation beam into blood components cause the formation of electrically charged particles or secondary electrons.

These electrons damage the DNA of lymphocytes either by direct interaction or by reacting initicially with cell water to form free radicals. The damaged lymphocytes are unable to proliferate in the host and therefore cannot mediate transfusion-associated GVHD.

Investigational Therapies for Graft-versus-host Disease

There are a large number of clinical trials either ongoing or recently completed in the investigation of graft-versus-host disease treatment and prevention.

Graft-versus-tumor Effect

Graft-versus-tumor effect (GVT) or "graft versus leukemia" effect is the beneficial aspect of the Graft-versus-Host phenomenon. For example, HSCT patients with either acute and in particular chronic graft-versus-host disease after an allogeneic transplant tend to have a lower risk of cancer relapse.

This is due to a therapeutic immune reaction of the grafted donor T lymphocytes against the diseased bone marrow of the recipient. This lower rate of relapse accounts for the increased success rate of allogeneic transplants, compared to transplants from identical twins, and indicates that allogeneic HSCT is a form of immunotherapy. GVT is the major benefit of transplants that do not employ the highest immuno-suppressive regimens.

Graft versus tumor is mainly beneficial in diseases with slow progress, e.g. chronic leukemia, low-grade lymphoma, and some cases multiple myeloma. However, it is less effective in rapidly growing acute leukemias.

If cancer relapses after HSCT, another transplant can be performed, infusing the patient with a greater quantity of donor white blood cells.

Oral Carcinoma

Patients after HSCT are at a higher risk for oral carcinoma. Post-HSCT oral cancer may have more aggressive behavior with poorer prognosis, when compared to oral cancer in non-HSCT patients.

Prognosis

Prognosis in HSCT varies widely dependent upon disease type, stage, stem cell source, HLA-matched status (for allogeneic HCST) and conditioning regimen. A transplant offers a chance for cure or long-term remission if the inherent complications of graft versus host disease, immuno-suppressive treatments and the spectrum of opportunistic infections can be survived. In recent years, survival rates have been gradually improving across almost all populations and sub-populations receiving transplants.

Mortality for allogeneic stem cell transplantation can be estimated using the prediction model created by Sorror et al., using the Hematopoietic Cell Transplantation-Specific Comorbidity Index (HCT-CI). The HCT-CI was derived and validated by investigators at the Fred Hutchinson Cancer Research Center (Seattle, WA). The HCT-CI modifies and adds to a well-validated comorbidity index, the Charlson Comorbidity Index (CCI) (Charlson et al.) The CCI was previously applied to patients undergoing allogeneic HCT but appears to provide less survival prediction and discrimination than the HCT-CI scoring system.

Risks to Donor

The risks of a complication depend on patient characteristics, health care providers and the apheresis procedure, and the colony-stimulating factor used (G-CSF, GM-CSF). G-CSF drugs include Filgrastim (Neupogen, Neulasta), and lenograstim (Graslopin).

Drug Risks

Filgrastim is typically dosed in the 10 microgram/kg level for 4–5 days during the harvesting of stem cells. The documented adverse effects of filgrastim include splenic rupture (indicated by left upper abdominal or shoulder pain, risk 1 in 40000), Adult respiratory distress syndrome (ARDS), alveolar hemorrage, and allergic reactions (usually expressed in first 30 minutes, risk 1 in 300). In addition, platelet and hemoglobin levels dip post-procedure, not returning to normal until one month.

The question of whether patients over 65 react the same as patients under 65 has not been sufficiently examined. Coagulation issues and inflammation of atherosclerotic plaques are known to occur as a result of G-CSF injection. G-CSF has also been described to induce genetic changes in mononuclear cells of normal donors. There is evidence that myelodysplasia (MDS) or acute myeloid leukaemia (AML) can be induced by GCSF in susceptible individuals.

Access Risks

Blood was drawn peripherally in a majority of patients, but a central line to jugular/subclavian/femoral veins may be used in 16% of women and 4% of men. Adverse reactions during apheresis were experienced in 20% of women and 8% of men, these adverse events primarily consisted of numbness/tingling, multiple line attempts, and nausea.

Clinical Observations

A study involving 2408 donors (18–60 years) indicated that bone pain (primarily back and hips) as a result of filgrastim treatment is observed in 80% of donors by day 4 post-injection. This pain responded to acetaminophen or ibuprofen in 65% of donors and was characterized as mild to moderate in 80% of donors and severe in 10%. Bone pain receded post-donation to 26% of patients 2 days post-donation, 6% of patients one week post-donation, and <2% 1 year post-donation. Donation is not recommended for those with a history of back pain. Other symptoms observed in more than 40% of donors include myalgia, headache, fatigue, and insomnia. These symptoms all returned to baseline 1 month post-donation, except for some cases of persistent fatigue in 3% of donors.. In one metastudy that incorporated data from 377 donors, 44% of patients reported having adverse side effects after peripheral blood HSCT. Side effects included pain prior to the collection procedure as a result of GCSF injections, post-procedural generalized skeletal pain, fatigue and reduced energy.

Severe Reactions

A study that surveyed 2408 donors found that serious adverse events (requiring prolonged hospitalization) occurred in 15 donors (at a rate of 0.6%), although none of these events were fatal. Donors were not observed to have higher than normal rates of cancer with up to 4–8 years of follow up. One study based on a survey of medical teams covered approximately 24,000 peripheral blood HSCT cases between 1993 and 2005, and found a serious cardiovascular adverse reaction rate of about 1 in 1500. This study reported a cardiovascular-related fatality risk within the first 30 days HSCT of about 2 in 10000. For this same group, severe cardiovascular events were observed with a rate of about 1 in 1500. The most common severe adverse reactions were pulmonary edema/deep vein thrombosis, splenic rupture, and myocardial infarction. Haematological malignancy induction was comparable to that observed in the general population with only 15 reported cases within 4 years...

Donor Registration and Recruitment

At the end of 2010, 14.9 million people had registered their willingness to be a bone marrow donor with one of the 64 registries from 45 countries participating in Bone Marrow Donors Worldwide. 12.2 million of these registered donors had been ABDR typed, allowing easy matching. A further 453,000 cord blood units had been received by one of 44 cord blood units from 26 countries participating. The highest total number of bone marrow donors registered were those from the USA (6.4 million), and the highest number per capita were those from Cyprus (10.6% of the population).

Experimental HIV Treatment

In 2007, a team of doctors in Berlin, Germany, including Gero Hütter, performed a stem cell transplant for a leukaemia patient who was also HIV-positive. From 60 matching donors, they selected a [CCR5]-Ä32 homozygous individual with two genetic copies of a rare variant of a cell surface receptor. This genetic trait confers resistance to HIV infection by blocking attachment of HIV to the cell. Roughly one in 1000 people of European ancestry have this inherited mutation, but it is rarer in other populations. The transplant was repeated a year later after a relapse. Over three years after the initial transplant and despite discontinuing antiretroviral therapy, researchers cannot detect

HIV in the transplant recipient's blood or in various biopsies. Levels of HIV-specific antibodies have also declined, leading to speculation that the patient may have been functionally cured of HIV. However, scientists emphasise that this is an unusual case. Potentially fatal transplant complications (the "Berlin patient" suffered from graft-versus-host disease and leukoencephalopathy) mean that the procedure could not be performed in others with HIV, even if sufficient numbers of suitable donors were found.

Potential Treatments

Brain Damage

Stroke and traumatic brain injury lead to cell death, characterized by a loss of neurons and oligodendrocytes within the brain. Healthy adult brains contain neural stem cells which divide to maintain general stem cell numbers, or become progenitor cells. In healthy adult animals, progenitor cells migrate within the brain and function primarily to maintain neuron populations for olfaction (the sense of smell).

Interestingly, in pregnancy and after injury, this system appears to be regulated by growth factors and can increase the rate at which new brain matter is formed. Although the reparative process appears to initiate following trauma to the brain, substantial recovery is rarely observed in adults, suggesting a lack of robustness. Stem cells may also be used to treat brain degeneration, such as in Parkinson's and Alzheimer's disease.

Cancer

The development of gene therapy strategies for treatment of intra-cranial tumours offers much promise, and has shown to be successful in the treatment of some dogs; although research in this area is still at an early stage. Using conventional techniques, brain cancer is difficult to treat because it spreads so rapidly. Researchers at the Harvard Medical School transplanted human neural stem cells into the brain of rodents that received intracranial tumours. Within days, the cells migrated into the cancerous area and produced cytosine deaminase, an enzyme that converts a non-toxic pro-drug into a chemotheraputic agent. As a result, the injected substance was able to reduce the tumor mass by 81 percent.

The stem cells neither differentiated nor turned tumorigenic. Some researchers believe that the key to finding a cure for cancer is to inhibit proliferation of cancer stem cells. Accordingly, current cancer treatments are designed to kill cancer cells. However, conventional chemotherapy treatments cannot discriminate between cancerous cells and others. Stem cell therapies may serve as potential treatments for cancer. Research on treating Lymphoma using adult stem cells is underway and has had human trials. Essentially, chemotherapy is used to completely destroy the patients own lymphocytes, and stem cells injected, eventually replacing the immune system of the patient with that of the healthy donor.

Spinal Cord Injury

A team of Korean researchers reported on November 25, 2003, that they had transplanted multipotent adult stem cells from umbilical cord blood to a patient suffering from a spinal cord injury and that following the procedure, she could walk on her own, without difficulty. The patient had not been able to stand up for roughly 19 years. For the unprecedented clinical test, the scientists isolated adult stem cells from umbilical cord blood and then injected them into the damaged part of the spinal cord.

According to the October 7, 2005 issue of The Week, University of California, Irvine researchers transplanted multipotent human fetal-derived neural stem cells into paralyzed mice, resulting in locomotor improvements four months later. The observed recovery was associated with differentiation of transplanted cells into new neurons and oligodendrocytes- the latter of which forms the myelin sheath around axons of the central nervous system, thus insulating neural impulses and facilitating communication with the brain.

In January 2005, researchers at the University of Wisconsin–Madison differentiated human blastocyst stem cells into neural stem cells, then into pre-mature motor neurons, and finally into spinal motor neurons, the cell type that, in the human body, transmits messages from the brain to the spinal cord and subsequently mediates motor function in the periphery. The newly generated motor neurons exhibited electrical activity, the signature action of neurons. Lead researcher Su-Chun Zhang described the process as "[teaching] the blastocyst stem cells to change step by step, where each step has different conditions and a strict window of time."

Transformation of blastocyst stem cells into motor neurons had eluded researchers for decades. While Zhang's findings were a significant contribution to the field, the ability of transplanted neural cells to establish communication with neighboring cells remains unclear. Accordingly, studies using chicken embryos as a model organism can be an effective proof-of-concept experiment. If functional, the new cells could be used to treat diseases like Lou Gehrig's disease, muscular dystrophy, and spinal cord injuries.

Heart Damage

Several clinical trials targeting heart disease have shown that adult stem cell therapy is safe, effective, and equally efficient in treating old and recent infarcts. Adult stem cell therapy for treating heart disease was commercially available in at least five continents at the last count (2007).

Possible mechanisms of recovery include:

- Generation of heart muscle cells
- Stimulation of growth of new blood vessels to repopulate damaged heart tissue
- Secretion of growth factors
- Assistance via some other mechanism

It may be possible to have adult bone marrow cells differentiate into heart muscle cells.

Haematopoiesis (Blood Cell Formation)

The specificity of the human immune cell repertoire is what allows the human body to defend itself from rapidly adapting antigens. However, the immune system is vulnerable to degradation upon the pathogenesis of disease, and because of the critical role that it plays in overall defense, its degradation is often fatal to the organism as a whole. Diseases of hematopoietic cells are called hematopathology. The specificity of the immune cells is what allows recognition of foreign antigens, causing further challenges in the treatment of immune disease. Identical matches between donor and recipient must be made for successful transplantation treatments, but matches are uncommon, even between first-degree relatives.

Research using both hematopoietic adult stem cells and embryonic stem cells has provided insight into the possible mechanisms and methods of treatment for many of these ailments.

Fully mature human red blood cells may be generated ex vivo by hematopoietic stem cells (HSCs), which are precursors of red blood cells. In this process, HSCs are grown together with stromal cells, creating an environment that mimics the conditions of bone marrow, the natural site of red blood cell growth. Erythropoietin, a growth factor, is added, coaxing the stem cells to complete terminal differentiation into red blood cells. Further research into this technique should have potential benefits to gene therapy, blood transfusion, and topical medicine.

Baldness

Hair follicles also contain stem cells, and some researchers predict research on these follicle stem cells may lead to successes in treating baldness through an activation of the stem cells progenitor cells. This treatment is expected to work by activating already existing stem cells on the scalp. Later treatments may be able to simply signal follicle stem cells to give off chemical signals to nearby follicle cells which have shrunk during the aging process, which in turn respond to these signals by regenerating and once again making healthy hair.

Most recently, Dr. Aeron Potter of the University of California has claimed that stem cell therapy led to a significant and visible improvement in follicular hair growth· Results from his experiments are under review by the journal Science (journal).

Missing Teeth

In 2004, scientists at King's College London discovered a way to cultivate a complete tooth in mice and were able to grow them stand-alone in the laboratory. Researchers are confident that this technology can be used to grow live teeth in human patients.

In theory, stem cells taken from the patient could be coaxed in the lab into turning into a tooth bud which, when implanted in the gums, will give rise to a new tooth, and would be expected to grow within two months. It will fuse with the jawbone and release chemicals that encourage nerves and blood vessels to connect with it. The process is similar to what happens when humans grow their original adult teeth.

Many challenges remain, however, before stem cells could be a choice for the replacement of missing teeth in the future.

Deafness

Heller has reported success in re-growing cochlea hair cells with the use of embryonic stem cells.

Blindness and Vision Impairment

Since 2003, researchers have successfully transplanted corneal stem cells into damaged eyes to restore vision. "Sheets of retinal cells used by the team are harvested from aborted fetuses, which some people find objectionable." When these sheets are transplanted over the damaged cornea, the stem cells stimulate renewed repair, eventually restore vision.

The latest such development was in June 2005, when researchers at the Queen Victoria Hospital of Sussex, England were able to restore the sight of forty patients using the same technique. The group, led by Dr. Sheraz Daya, was able to successfully use adult stem cells obtained from the patient, a relative, or even a cadaver. Further rounds of trials are ongoing.

In April 2005, doctors in the UK transplanted corneal stem cells from an organ donor to the cornea of Deborah Catlyn, a woman who was blinded in one eye when acid was thrown in her eye at a nightclub. The cornea, which is the transparent window of the eye, is a particularly suitable site for transplants. In fact, the first successful human transplant was a cornea transplant. The absence of blood vessels within the cornea makes this area a relatively easy target for transplantation. The majority of corneal transplants carried out today are due to a degenerative disease called keratoconus. The University Hospital of New Jersey reports that the success rate for growth of new cells from transplanted stem cells varies from 25 percent to 70 percent.

In 2009, researchers at the University of Pittsburgh Medical center demonstrated that stem cells collected from human corneas can restore transparency without provoking a rejection response in mice with corneal damage.

Amyotrophic Lateral Sclerosis

Stem cells have resulted in significant locomotor improvements in rats with an Amyotrophic lateral sclerosis-like disease. In a rodent model that closely mimics the human form of ALS, animals were injected with a virus to kill the spinal cord motor nerves which mediate movement. Animals subsequently received stem cells in the spinal cord. Transplanted cells migrated to the sites of injury, contributed to regeneration of the ablated nerve cells, and restored locomotor function.

Graft vs. Host Disease and Crohn's Disease

Phase III clinical trials expected to end in second-quarter 2008 were conducted by Osiris Therapeutics using their in-development product Prochymal, derived from adult bone marrow. The target disorders of this therapeutic are graft-versus-host disease and Crohn's disease.

Neural and Behavioral Birth Defects

A team of researchers led by Prof. Joseph Yanai were able to reverse learning deficits in the offspring of pregnant mice who were exposed to heroin and the pesticide organophosphate. This was done by direct neural stem cell transplantation into the brains of the offspring. The recovery was almost 100 percent, as shown both in behavioral tests and objective brain chemistry tests. Behavioral tests and learning scores of the treated mice showed rapid improvement after treatment, providing results that rivaled non-treated mice. On the molecular level, brain chemistry of the treated animals was also restored to normal. Through the work, which was supported by the US National Institutes of Health, the US-Israel Binational Science Foundation and the Israel anti-drug authorities, the researchers discovered that the stem cells worked even in cases where most of the cells died out in the host brain.

The scientists found that before they die the neural stem cells succeed in inducing the host brain to produce large numbers of stem cells which repair the damage. These findings, which answered a major question in the stem cell research community, were published in January 2008 in the leading journal, Molecular Psychiatry. Scientists are now developing procedures to administer the neural stem cells in the least invasive way possible - probably via blood vessels, making therapy practical and clinically feasible. Researchers also plan to work on developing methods to take cells from the patient's own body, turn them into stem cells, and then transplant them back into the patient's blood via the blood stream. Aside from decreasing the chances of immunological rejection, the approach will also eliminate the controversial ethical issues involved in the use of stem cells from human embryos.

Diabetes

Diabetes patients lose the function of insulin-producing beta cells within the pancreas. Human embryonic stem cells may be grown in cell culture and stimulated to form insulin-producing cells that can be transplanted into the patient.

However, clinical success is highly dependent on the development of the following procedures:

- Transplanted cells should proliferate
- Transplanted cells should differentiate in a site-specific manner
- Transplanted cells should survive in the recipient (prevention of transplant rejection)
- Transplanted cells should integrate within the targeted tissue
- Transplanted cells should integrate into the host circuitry and restore function

Orthopaedics

Clinical case reports in the treatment of orthopaedic conditions have been reported. To date, the focus in the literature for musculoskeletal care appears to be on mesenchymal stem cells. Centeno et al. have published MRI evidence of increased cartilage and meniscus volume in individual human subjects. The results of trials that include a large number of subjects, are yet to be published. However, a published safety study conducted in a group of 227 patients over a 3-4 year period shows adequate safety and minimal complications associated with mesenchymal cell transplantation.

Wakitani has also published a small case series of nine defects in five knees involving surgical transplantation of mesenchymal stem cells with coverage of the treated chondral defects.

Wound Healing

Stem cells can also be used to stimulate the growth of human tissues. In an adult, wounded tissue is most often replaced by scar tissue, which is characterized in the skin by disorganized collagen structure, loss of hair follicles and irregular vascular structure. In the case of wounded fetal tissue, however, wounded tissue is replaced with normal tissue through the activity of stem cells. A possible method for tissue regeneration in adults is to place adult stem cell "seeds" inside a tissue bed "soil" in a wound bed and allow the stem cells to stimulate differentiation in the tissue bed cells. This method elicits a regenerative response more similar to fetal wound-healing than adult scar tissue formation. Researchers are still investigating different aspects of the "soil" tissue that are conducive to regeneration.

Infertility

Culture of human embryonic stem cells in mitotically inactivated porcine ovarian fibroblasts (POF) causes differentiation into germ cells (precursor cells of oocytes and spermatozoa), as evidenced by gene expression analysis.

Human embryonic stem cells have been stimulated to form Spermatozoon-like cells, yet still slightly damaged or malformed. It could potentially treat azoospermia.

7

Gene Networks

A gene regulatory network or genetic regulatory network (GRN) is a collection of DNA segments in a cell which interact with each other indirectly (through their RNA and protein expression products) and with other substances in the cell, thereby governing the rates at which genes in the network are transcribed into mRNA. In general, each mRNA molecule goes on to make a specific protein (or set of proteins). In some cases this protein will be structural, and will accumulate at the cell membrane or within the cell to give it particular structural properties.

In other cases the protein will be an enzyme, i.e., a micro-machine that catalyses a certain reaction, such as the breakdown of a food source or toxin. Some proteins though serve only to activate other genes, and these are the transcription factors that are the main players in regulatory networks or cascades. By binding to the promoter region at the start of other genes they turn them on, initiating the production of another protein, and so on. Some transcription factors are inhibitory.

In single-celled organisms, regulatory networks respond to the external environment, optimising the cell at a given time for survival in this environment. Thus a yeast cell, finding itself in a sugar solution, will turn on genes to make enzymes that process the sugar to alcohol. This process, which we associate with wine-making, is how the yeast cell makes its living, gaining energy to multiply, which under normal circumstances would enhance its survival prospects. In multicellular animals the same principle has been put in the service of gene cascades that control body-shape. Each time a cell divides, two cells result which, although they contain the same genome in full, can differ in which genes are turned on and making proteins. Sometimes a 'self-sustaining feedback loop' ensures that a cell maintains its identity and passes it on. Less understood is the mechanism of epigenetics by which chromatin modification may provide cellular memory by blocking or allowing transcription. A major feature of multicellular animals is the use of morphogen gradients, which in effect provide a positioning system that tells a cell where in the body it is, and hence what sort of cell to become. A gene that is turned on in one cell may make a product that leaves the cell and diffuses through adjacent cells, entering them and turning on genes only when it is present above a certain threshold level.

These cells are thus induced into a new fate, and may even generate other morphogens that signal back to the original cell. Over longer distances morphogens may use the active process of signal transduction. Such signalling controls embryogenesis, the building of a body plan from scratch through a series of sequential steps.

They also control maintain adult bodies through feedback processes, and the loss of such feedback because of a mutation can be responsible for the cell proliferation that is seen in cancer. In parallel with this process of building structure, the gene cascade turns on genes that make structural proteins that give each cell the physical properties it needs. It has been suggested that, because biological molecular interactions are intrinsically stochastic, gene networks are the result of cellular processes and not their cause (i.e. Cellular Darwinism). However, recent experimental evidence has favored the attractor view of cell fates.

Genes have sometimes been regarded as nodes in a network, with inputs being proteins such as transcription factors, and outputs being the level of gene expression. The node itself performs a function, and the operation of these functions have been interpreted as performing a kind of information processing within cell and determine cellular behavior.

Gene networks can also be constructed without formulating an explicit causal model. This is often the case when assembling networks from large expression data sets. Covariation and correlation of expression is computed across a large sample of cases and measurements (often transcriptome or proteome data). The source of variation can be either experimental or natural (observational).

There are several ways to construct gene expression networks, but one common approach is to compute a matrix of all pair-wise correlations of expression across conditions, time points, or individuals and convert the matrix (after thresholding at some cut-off value) into a graphical representation in which nodes represent genes, transcripts, or proteins and edges connecting these nodes represent the strength of association.

Overview

At one level, biological cells can be thought of as "partially-mixed bags" of biological chemicals – in the discussion of gene regulatory networks, these chemicals are mostly the mRNAs and proteins that arise from gene expression. These mRNA and proteins interact with each other with various degrees of specificity. Some diffuse around the cell. Others are bound to cell membranes, interacting with molecules in the environment. Still others pass through cell membranes and mediate long range signals to other cells in a multi-cellular organism. These molecules and their interactions comprise a gene regulatory network. A typical gene regulatory network looks something like this:

The nodes of this network are proteins, their corresponding mRNAs, and protein/protein complexes. Nodes that are depicted as lying along vertical lines are associated with the cell/environment interfaces, while the others are free-floating and diffusible. Implied are genes, the DNA sequences which are transcribed into the mRNAs that translate into proteins. Edges between nodes represent individual molecular reactions, the protein/protein and protein/mRNA interactions through which the products of one gene affect those of another, though the lack of experimentally obtained information often implies that some reactions are not modeled at such a fine level of detail. These interactions can be inductive (the arrowheads), with an increase in the concentration of one leading to an increase in the other, or inhibitory (the filled circles), with an increase in one leading to a decrease in the other. A series of edges indicates a chain of such dependences, with cycles corresponding to feedback loops. The network structure is an abstraction of the system's chemical dynamics, describing the manifold ways in which one substance affects all the others to which it is connected. In practice, such GRNs are inferred from the biological literature on a given system and represent a distillation of the collective knowledge about a set of related biochemical reactions.

Genes can be viewed as nodes in the network, with input being proteins such as transcription factors, and outputs being the level of gene expression. The node itself can also be viewed as a function which can be obtained by combining basic functions upon the inputs (in the Boolean network described below these are Boolean functions, typically AND, OR, and NOT). These functions have been interpreted as performing a kind of information processing within the cell, which determines cellular behavior. The basic drivers within cells are concentrations of some proteins, which determine both spatial (location within the cell or tissue) and temporal (cell cycle or developmental stage) coordinates of the cell, as a kind of "cellular memory". The gene networks are only beginning to be understood, and it is a next step for biology to attempt to deduce the functions for each gene "node", to help understand the behavior of the system in increasing levels of complexity, from gene to signaling pathway, cell or tissue level. Mathematical models of GRNs have been developed to capture the behavior of the system being modeled, and in some cases generate predictions corresponding with experimental observations. In some other cases, models have proven to make accurate novel predictions, which can be tested experimentally, thus suggesting new approaches to explore in an experiment that sometimes wouldn't be considered in the design of the protocol of an experimental laboratory. The most common modeling technique involves the use of coupled ordinary differential equations (ODEs). Several other promising modeling techniques have been used, including Boolean networks, Petri nets, Bayesian networks, graphical Gaussian models, Stochastic, and Process Calculi. Conversely, techniques have been proposed for generating models of GRNs that best explain a set of time series observations.

Modelling

Boolean Network

The following example illustrates how a Boolean network can model a GRN together with its gene products (the outputs) and the substances from the environment that affect it (the inputs). Stuart Kauffman was amongst the first biologists to use the metaphor of Boolean networks to model genetic regulatory networks.

1. Each gene, each input, and each output is represented by a node in a directed graph in which there is an arrow from one node to another if and only if there is a causal link between the two nodes.
2. Each node in the graph can be in one of two states: on or off.
3. For a gene, "on" corresponds to the gene being expressed; for inputs and outputs, "on" corresponds to the substance being present.
4. Time is viewed as proceeding in discrete steps. At each step, the new state of a node is a Boolean function of the prior states of the nodes with arrows pointing towards it.

The validity of the model can be tested by comparing simulation results with time series observations.

Continuous Networks

Continuous network models of GRNs are an extension of the boolean networks described above. Nodes still represent genes and connections between them regulatory influences on gene expression. Genes in biological systems display a continuous range of activity levels and it has been argued that using a continuous representation captures several properties of gene regulatory networks not present in the Boolean model. Formally most of these approaches are similar to an artificial neural network, as inputs to a node are summed up and the result serves as input to a sigmoid function, e.g., but proteins do often control gene expression in a synergistic, i.e. non-linear, way. However there is now a continuous network model that allows grouping of inputs to a node thus realizing another level of regulation. This model is formally closer to a higher order recurrent neural network. The same model has also been used to mimic the evolution of cellular differentiation and even multicellular morphogenesis.

Stochastic Gene Networks

Recent experimental results have demonstrated that gene expression is a stochastic process. Thus, many authors are now using the stochastic formalism, after the work by. Works on single gene expression and small synthetic genetic networks, such as the genetic toggle switch of Tim Gardner and Jim Collins, provided additional experimental data on the phenotypic variability and the stochastic nature of gene expression. The first versions of stochastic models of gene expression involved only instantaneous reactions and were driven by the Gillespie algorithm.

Since some processes, such as gene transcription, involve many reactions and could not be correctly modeled as an instantaneous reaction in a single step, it was proposed to model these reactions as single step multiple delayed reactions in order to account for the time it takes for the entire process to be complete. From here, a set of reactions were proposed that allow generating GRNs. These are then simulated using a modified version of the Gillespie algorithm, that can simulate multiple time delayed reactions (chemical reactions where each of the products is provided a time delay that determines when will it be released in the system as a "finished product").

For example, basic transcription of a gene can be represented by the following single-step reaction (RNAP is the RNA polymerase, RBS is the RNA ribosome binding site, and Pro_i is the promoter region of gene i):

$$RNAP + \mathrm{Pr}\,o_i \xrightarrow{k_i, bas} \mathrm{Pr}\,o_i\left(\tau_i^1\right) + RBS_i\left(\tau_i^1\right) + RNAp\left(\tau_I^2\right)$$

A recent work proposed a simulator (SGNSim, Stochastic Gene Networks Simulator), that can model GRNs where transcription and translation are modeled as multiple time delayed events and its dynamics is driven by a stochastic simulation algorithm (SSA) able to deal with multiple time delayed events. The time delays can be drawn from several distributions and the reaction rates from complex functions or from physical parameters. SGNSim can generate ensembles of GRNs within a set of user-defined parameters, such as topology. It can also be used to model specific GRNs and systems of chemical reactions. Genetic perturbations such as gene deletions, gene over-expression, insertions, frame shift mutations can also be modeled as well.

The GRN is created from a graph with the desired topology, imposing in-degree and out-degree distributions. Gene promoter activities are affected by other genes expression products that act as inputs, in the form of monomers or combined into multimers and set as direct or indirect.

Next, each direct input is assigned to an operator site and different transcription factors can be allowed, or not, to compete for the same operator site, while indirect inputs are given a target. Finally, a function is assigned to each gene, defining the gene's response to a combination of transcription factors (promoter state). The transfer functions (that is, how genes respond to a combination of inputs) can be assigned to each combination of promoter states as desired.

In other recent work, multiscale models of gene regulatory networks have been developed that focus on synthetic biology applications. Simulations have been used that model all biomolecular interactions in transcription, translation, regulation, and induction of gene regulatory networks, guiding the design of synthetic systems.

Prediction

Other work has focused on predicting the gene expression levels in a gene regulatory network. The approaches used to model gene regulatory networks have been constrained to be interpretable and, as a result, are generally simplified versions of the network. For example, Boolean networks have been used due to their simplicity and ability to handle noisy data but lose data information by having a binary representation of the genes. Also, artificial neural networks omit using a hidden layer so that they can be interpreted, losing the ability to model higher order correlations in the data. Using a model that is not constrained to be interpretable, a more accurate model can be produced. Being able to predict gene expressions more accurately provides a way to explore how drugs affect a system of genes as well as for finding which genes are interrelated in a process. This has been encouraged by the DREAM competition which promotes a competition for the best prediction algorithms. Some other recent work has used artificial neural networks with a hidden layer.

Network Connectivity

Empirical data indicate that biological gene networks are sparsely connected, and that the average number of upstream-regulators per gene is less than two. Theoretical results show that selection for robust gene networks will favor minimally complex, more sparsely connected, networks. These results suggest that a sparse, minimally connected, genetic architecture may be a fundamental design constraint shaping the evolution of gene network complexity.

Techniques and Tools

The following experimental techniques are used to measure gene expression and are listed in roughly chronological order, starting with the older, more established technologies. They are divided into two groups based on their degree of multiplexity.

- Low-to-mid-plex techniques:
  - Reporter gene
  - Northern blot
  - Western blot
  - Fluorescent in situ hybridization
  - Reverse transcription PCR
  - Digital counting of single transcript molecules, see NanoString Technologies
- Higher-plex techniques:
  - SAGE
  - DNA microarray
  - Tiling array
  - RNA-Seq

Transcriptional Noise

A primary cause of the variability (noise) in gene expression occurring between cells in isogenic populations. A major source of transcriptional noise is likely to be transcriptional bursting although other sources of heterogeneity, such as unequal separation of cell contents at mitosis are also likely to contribute considerably. Bursting transcription, as opposed to simple probabilistic models of transcription, reflects multiple states of gene activity, with fluctuations between states separated by irregular intervals, generating uneven protein expression between cells. Noise in gene expression can have tremendous consequences on cell behaviour, and must be mitigated or integrated. In certain contexts, such as the survival of microbes in rapidly changing stressful environments, or several types of scattered differentiation, the variability may be essential. Variability also impacts upon the effectiveness of clinical treatment, with resistance of bacteria to antibiotics demonstrably caused by non-genetic differences. Variability in gene expression may also contribute to resistance of sub-populations of cancer cells to chemotherapy.

Transcriptional Bursting

Transcriptional bursting, also known as transcriptional pulsing, is a fundamental property of genes from bacteria to humans. Transcription of genes, the process which transforms the stable code written in DNA into the mobile RNA message can occur in "bursts" or "pulses". This phenomenon has recently come to light with the advent of new technologies, such as MS2 tagging, to detect RNA production in single cells, allowing precise measurements of RNA number, or RNA appearance at the gene. Other, more widespread techniques, such as Northern Blotting, Microarrays, RT-PCR and RNA-Seq, measure bulk RNA levels from homogenous population extracts. These techniques lose dynamic information from individual cells, and give the impression transcription is a continuous smooth process. The reality is that transcription is irregular, with strong periods of activity, interspersed by long periods of inactivity. Averaged over millions of cells, this appears continuous. But at the individual cell level, there is considerable variability, and for most genes, very little activity at any one time.

Bursting may result from the stochastic nature of biochemical events superimposed upon a 2 or more step fluctuation. In its most simple form, the gene can exist in 2 states, one where activity is negligible and one where there is a certain probability of activation. Only in the second state does transcription readily occur. Whilst the nuclear and signaling landscapes of complex eukaryotic nuclei are likely to favour more than two simple states- for example, there are over twenty post-translational modifications of nucleosomes known, this simple two step model perhaps provides a reasonable intellectual framework for understanding the changing probabilities affecting transcription. It seems likely that some rudimentary eukaryotes have genes which do not show bursting. The genes are always in the permissive state, with a simple probability describing the numbers of RNAs generated. What do the repressive and permissive states represent? An attractive idea is that the repressed state is a closed chromatin conformation whilst the permissive state is an open one. Another hypothesis is that the fluctuations reflect transition between bound pre-initiation complexes (permissive) and dissociated ones (restrictive). Bursts may also result from bursty signalling, cell cycle effects or movement of chromatin to and from transcription factories.

The bursting phenomenon, as opposed to simple probabilistic models of transcription, can account for the high variability in gene expression occurring between cells in isogenic populations. This variability in turn can have tremendous consequences on cell behaviour, and must be mitigated or integrated. In certain contexts, such as the survival of microbes in rapidly changing stressful environments, or several types of scattered differentiation, the variability may be essential. Variability also impacts upon the effectiveness of clinical treatment, with resistance of bacteria to antibiotics demonstrably caused by non-genetic differences. Variability in gene expression may also contribute to resistance of sub-populations of cancer cells to chemotherapy.

Bookmarking

In genetics and epigenetics, bookmarking is a biological phenomenon believed to function as an epigenetic mechanism for transmitting cellular memory of the pattern of gene expression in a cell, throughout mitosis, to its daughter cells. This is vital for maintaining the phenotype in a lineage of cells so that, for example, liver cells divide into liver cells and not some other cell type.

It is characterized by non-compaction of some gene promoters during mitosis. In terms of mechanism, it is believed that:

- at some point prior to the onset of mitosis, the promoters of genes that exist in a transcription-competent state become "marked" in some way,
- this "mark" persists both during and after mitosis, and
- the marking transmits gene expression memory by preventing the mitotic compaction of DNA at this locus, or by facilitating reassembly of transcription complexes on the promoter, or both.

In some cases, bookmarking is mediated by binding of specific factors to the promoter prior to onset of mitosis, but in other cases could be mediated by patterns of histone modification or presence of histone variants that are characteristic of active genes, and which are believed to persist throughout mitosis.

In the case of specific genes, for example, the stress-inducible hsp70 gene, bookmarking may also function as a mechanism for ensuring that the gene can be transcribed early in G1 phase if a stress were to occur at that time. If this gene promoter were compacted it would take time to de-compact in G1, during which time the cell would be unable to transcribe this cytoprotective gene, leaving it vulnerable to stress-induced cell death. In this case, bookmarking appears to be important for cell survival.

Expressed Sequence Tag

An expressed sequence tag or EST is a short sub-sequence of a cDNA sequence. They may be used to identify gene transcripts, and are instrumental in gene discovery and gene sequence determination. The identification of ESTs has proceeded rapidly, with approximately 65.9 million ESTs now available in public databases (e.g. GenBank 18 June 2010, all species).

An EST results from one-shot sequencing of a cloned mRNA (i.e. several hundred base pairs of sequence starting from an end of a cDNA). The cDNAs used for EST generation are typically individual clones from a cDNA library. The resulting sequence is a relatively low quality fragment whose length is limited by current technology to approximately 500 to 800 nucleotides. Because these clones consist of DNA that is complementary to mRNA, the ESTs represent portions of expressed genes. They may be represented in databases as either cDNA/mRNA sequence or as the reverse complement of the mRNA, the template strand.

ESTs can be mapped to specific chromosome locations using physical mapping techniques, such as radiation hybrid mapping, Happy mapping, or FISH. Alternatively, if the genome of the organism that originated the EST has been sequenced, one can align the EST sequence to that genome using a computer.

The current understanding of the human set of genes (as of 2006) includes the existence of thousands of genes based solely on EST evidence. In this respect, ESTs have become a tool to refine the predicted transcripts for those genes, which leads to the prediction of their protein products and ultimately their function. Moreover, the situation in which those ESTs are obtained (tissue, organ, disease state - e.g. cancer) gives information on the conditions in which the corresponding gene is acting. ESTs contain enough information to permit the design of precise probes for DNA microarrays that then can be used to determine the gene expression.

Some authors use the term "EST" to describe genes for which little or no further information exists besides the tag.

The significance of ESTs, their properties, methods to analyze EST dataset and their applications in different areas of biology have been reviewed by Nagaraj et al. (2007).

Sources of Data and Annotations

dbEST

dbEST is a division of Genbank established in 1992. As for GenBank, data in dbEST is directly submitted by laboratories worldwide and is not curated.

EST Contigs

Because of the way ESTs are sequenced, many distinct expressed sequence tags are often partial sequences that correspond to the same mRNA of an organism. In an effort to reduce the number of expressed sequence tags for downstream gene discovery analyses, several groups assembled expressed sequence tags into EST contigs. Example of resources that provide EST contigs include:

- TIGR gene indices
- Unigene
- STACK

Constructing EST contigs is not trivial and may yield artifacts (contigs that contain two distinct gene products). When the complete genome sequence of an organism is available and transcripts are annotated, it is possible to bypass contig assembly and directly match transcripts with ESTs. This approach is used in the TissueInfo system and makes it easy to link annotations in the genomic database to tissue information provided by EST data.

Tissue Information

High-throughput analyses of ESTs often encounter similar data management challenges. A first challenge is that tissue provenance of EST libraries is described in plain English in dbEST. This makes it difficult to write programs that can non ambiguously determine that two EST libraries were sequenced from the same tissue. Similarly, disease conditions for the tissue are not annotated in a computationally friendly manner. For instance, cancer origin of a library is often mixed with the tissue name (e.g., the tissue name "glioblastoma" indicates that the EST library was sequenced from brain tissue and the disease condition is cancer).

With the notable exception of cancer, the disease condition is often not recorded in dbEST entries. The TissueInfo project was started in 2000 to help with these challenges. The project provides curated data (updated daily) to disambiguate tissue origin and disease state (cancer/non cancer), offers a tissue ontology that links tissues and organs by "is part of" relationships (i.e., formalizes knowledge that hypothalamus is part of brain, and that brain is part of the central nervous system) and distributes open-source software for linking transcript annotations from sequenced genomes to tissue expression profiles calculated with data in dbEST.

Paramutation

In epigenetics, paramutation is an interaction between two alleles of a single locus, resulting in a heritable change of one allele that is induced by the other allele. Paramutation violates Mendel's first law, which states that in the process of the formation of the gametes (egg or sperm) the allelic pairs separate, one going to each gamete, and that each allele remains completely uninfluenced by the other. In paramutation an allele in one generation heritably affects the other allele in future generations, even if the allele causing the change is itself not transmitted. What may be transmitted are patterns of DNA methylation or RNAs such as piRNAs, siRNAs, miRNAs or other regulatory RNAs. Through proper breeding, paramutation can result in isogenic sibling plants with drastically different phenotypes.

Paramutation was first discovered and studied in maize (Zea mays) by R.A. Brink at the University of Wisconsin–Madison in the 1950s. Brink noticed that specific weakly expressed alleles of the red1 (r1) locus in maize, which encodes a transcription factor that confers red pigment to corn kernels, can heritably change specific strongly expressed alleles to a weaker expression state. The weaker expression state adopted by the changed allele is heritable and can, in turn, change the expression state of other active alleles in a process termed secondary paramutation. Brink showed that the influence of the paramutagenic allele could persist for many generations.

Interestingly, paramutation can result in a single allele of a gene controlling a spectrum of phenotypes. At r1 in maize, for example, the weaker expression state adopted by an allele following paramutation can range from completely colorless to nearly fully colored kernels. This is an exception to the general observation that continuous variation is controlled by many genes.

Allelic interactions similar to paramutation have since been reported in other organisms, including tomato, pea, and mice.

The molecular basis of paramutation is being unraveled, almost exclusively in maize. Paramutation may share common mechanisms with other epigenetic phenomena, such as gene silencing and genomic imprinting. In maize, paramutation seems to share many traits with the well understood RNA-directed DNA-methylation pathway in Arabidopsis thaliana, even though it has never been observed in the famous model plant. Alleman (2006) reported that, in maize, "paramutation is RNA-directed. Stability of the chromatin states associated with paramutation and transposon silencing requires the mop1 gene, which encodes an RNA-dependent RNA polymerase." Exactly how the RNA produced by this polymerase causes paramutation in maize is not yet understood, but like other epigenetic changes, it involves the covalent modification of DNA and/or the DNA-bound histone proteins without changing the DNA sequence itself.

Sequence Profiling Tool

A sequence profiling tool in bioinformatics is a type of software that presents information related to a genetic sequence, gene name, or keyword input. Such tools generally take a query such as a DNA, RNA, or protein sequence or 'keyword' and search one or more databases for information related to that sequence. Summaries and aggregate results are provided in standardized format describing the information that would otherwise have required visits to many smaller sites or direct literature searches to compile. Many sequence profiling tools are software portals or gateways that simplify the process of finding information about a query in the large and growing number of bioinformatics databases. The access to these kinds of tools is either web based or locally downloadable executables.

8

Biological Functions and Evolution

As a first approximation, it is possible to view introns as unimportant sequences whose only function is to be removed from an unspliced precursor RNA in order to generate the functional mRNA, rRNA or tRNA product. However, it is now well-established that some introns themselves encode specific proteins or can be further processed after splicing to generate noncoding RNA molecules. Alternative splicing is widely used to generate multiple proteins from a single gene. Furthermore, some introns represent mobile genetic elements and may be regarded as examples of selfish DNA.

The biological origins of introns are obscure. After the initial discovery of introns in protein-coding genes of the eukaryotic nucleus, there was significant debate as to whether introns in modern-day organisms were inherited from a common ancient ancestor (termed the introns-early hypothesis), or whether they appeared in genes rather recently in the evolutionary process (termed the introns-late hypothesis). Another theory is that the spliceosome and the intron-exon structure of genes is a relic of the RNA world (the introns-first hypothesis). There is still considerable debate about the extent to which of these hypotheses is most correct. The popular consensus at the moment is that introns arose within the eukaryote lineage as selfish elements.

Early studies of genomic DNA sequences from a wide range of organisms show that the intron-exon structure of homologous genes in different organisms can vary widely. More recent studies of entire eukaryotic genomes have now shown that the lengths and density (introns/gene) of introns varies considerably between related species. For example, while the human genome contains an average of 8.4 introns/gene (139,418 in the genome), the unicellular fungus

Encephalitozoon cuniculi contains only 0.0075 introns/gene (15 introns in the genome). Since eukaryotes arose from a common ancestor (Common descent), there must have been extensive gain and/or loss of introns during evolutionary time. This process is thought to be subject to selection, with a tendency towards intron gain in larger species due to their smaller population sizes, and the converse in smaller (particularly unicellular) species. Biological factors also influence which genes in a genome lose or accumulate introns.

Alternative splicing of introns within a gene acts to introduce greater variability of protein sequences translated from a single gene, allowing multiple related proteins to be generated from a single gene and a single precursor mRNA transcript. The control of alternative RNA splicing is performed by complex network of signaling molecules that respond to a wide range of intracellular and extracellular signals.

Introns contain several short sequences that are important for efficient splicing, such as acceptor and donor sites at either end of the intron as well as a branch point site, which are required for proper splicing by the spliceosome. Some introns are known to enhance the expression of the gene that they are contained in by a process known as intron-mediated enhancement (IME).

Coding Regions

Coding regions are composed of codons, which are decoded and translated (in eukaryotes usually into one and in prokaryotes usually into several) proteins by the ribosome. Coding regions begin with the start codon and end with a stop codon. In general, the start codon is an AUG triplet and the stop codon is UAA, UAG, or UGA. The coding regions tend to be stabilised by internal base pairs, this impedes degradation. In addition to being protein-coding, portions of coding regions may serve as regulatory sequences in the pre-mRNA as exonic splicing enhancers or exonic splicing silencers.

The coding region of a gene, also known as the coding sequence or CDS, is that portion of a gene's DNA or RNA, composed of exons, that codes for protein.

The region is bounded nearer the 5' end by a start codon and nearer the 3' end with a stop codon. The coding region in mRNA is bounded by the five prime untranslated region and the three prime untranslated region, which are also parts of the exons. The coding region of an organism is sum total of the organism's genome that is composed of gene coding regions.

Coding Sequence Annotation

While identification of open reading frames within a DNA sequence is straightforward, identifying coding sequences is not, because the cell translates only a subset of all open reading frames to proteins. Currently CDS prediction uses sampling and sequencing of mRNA from cells, although there is still the problem of determining which parts of a given mRNA are actually translated to protein. CDS prediction is a subset of gene prediction, the latter also including prediction of DNA sequences that code not only for protein but also for other functional elements such as RNA genes and regulatory sequences.

Untranslated Regions

Untranslated regions (UTRs) are sections of the mRNA before the start codon and after the stop codon that are not translated, termed the five prime untranslated region (5' UTR) and three prime untranslated region (3' UTR), respectively. These regions are transcribed with the coding region and thus are exonic as they are present in the mature mRNA. Several roles in gene expression have been attributed to the untranslated regions, including mRNA stability, mRNA localization, and translational efficiency. The ability of a UTR to perform these functions depends on the sequence of the UTR and can differ between mRNAs.

The stability of mRNAs may be controlled by the 5' UTR and/or 3' UTR due to varying affinity for RNA degrading enzymes called ribonucleases and for ancillary proteins that can promote or inhibit RNA degradation.

Translational efficiency, including sometimes the complete inhibition of translation, can be controlled by UTRs. Proteins that bind to either the 3' or 5' UTR may affect translation by influencing the ribosome's ability to bind to the mRNA. MicroRNAs bound to the 3' UTR also may affect translational efficiency or mRNA stability.

Cytoplasmic localization of mRNA is thought to be a function of the 3' UTR. Proteins that are needed in a particular region of the cell can also be translated there; in such a case, the 3' UTR may contain sequences that allow the transcript to be localized to this region for translation.

Some of the elements contained in untranslated regions form a characteristic secondary structure when transcribed into RNA. These structural mRNA elements are involved in regulating the mRNA. Some, such as the SECIS element, are targets for proteins to bind. One class of mRNA element, the riboswitches, directly bind small molecules, changing their fold to modify levels of transcription or translation. In these cases, the mRNA regulates itself.

5' UTR

A messenger ribonucleic acid (mRNA) molecule codes for a protein through translation. The mRNA also contains regions that are not translated: in eukaryotes these include the 5' untranslated region, 3' untranslated region, 5' cap and poly-A tail.

The five prime untranslated region (5' UTR), can contain elements for controlling gene expression by way of regulatory elements. It begins at the transcription start site and ends one nucleotide (nt) before the start codon (usually AUG) of the coding region. In prokaryotes, the 5' UTR usually contains a ribosome binding site (RBS), also known as the Shine Dalgarno sequence (AGGAGGU).

The 5' UTR has a median length of ~150 nt in eukaryotes, but can be as long as several thousand bases. Some viruses and cellular genes have unusually long and structured 5' UTRs which may impact gene expression. On average, 3' UTR tend to be twice as long as the 5' UTR. In prokaryotic mRNAs the 5' UTR is normally shorter.

mRNA structure, approximately to scale for a human mRNA, where the median length of the 5'UTR is 170 nucleotides.

Several regulatory sequences may be found in the 5' UTR:

- Binding sites for proteins, that may affect the mRNA's stability or translation, for example iron responsive elements, that regulate gene expression in response to iron.
- Riboswitches.
- Sequences that promote or inhibit translation initiation.
- Introns within 5' UTRs have been linked to regulation of gene expression and mRNA export.

3' UTR

In molecular genetics, the three prime untranslated region (3' UTR) is a particular section of messenger RNA (mRNA). It follows the coding region.

An mRNA molecule codes for a protein through translation. The mRNA also contains regions that are not translated. In eukaryotes these regions are the cap, 5' untranslated region, 3' untranslated region, and polyA tail.

In prokaryotes mRNA structures have some differences (see mRNA) as do histone mRNAs. However, both have 3' UTRs.

Several regulatory sequences are found in the 3' UTR:

- A polyadenylation signal, usually AAUAAA, or a slight variant. This marks the site of cleavage of the transcript approximately 30 base pairs past the signal, followed by the addition of several hundred adenine residues (poly-A tail).
- Binding sites for proteins, that may affect the mRNA's stability or location in the cell, like SECIS elements (which direct the ribosome to translate the codon UGA as selenocysteines rather than as a stop

codon), or AU-rich elements (AREs), stretches consisting of mainly adenine and uracil nucleotides (which can either stabilize or destabilize the mRNA depending on the protein bound to it).

- Binding sites for miRNAs, a type of RNAi.

Poly(A) Tail

The 3' poly(A) tail is a long sequence of adenine nucleotides (often several hundred) added to the 3' end of the pre-mRNA. This tail promotes export from the nucleus and translation, and protects the mRNA from degradation.

Monocistronic Versus Polycistronic mRNA

An mRNA molecule is said to be monocistronic when it contains the genetic information to translate only a single protein chain (polypeptide). This is the case for most of the eukaryotic mRNAs. On the other hand, polycistronic mRNA carries several open reading frames (ORFs), each of which is translated into a polypeptide. These polypeptides usually have a related function (they often are the subunits composing a final complex protein) and their coding sequence is grouped and regulated together in an operon. Most of the mRNA found in bacteria and archea is polycistronic. Dicistronic or bicistronic is the term used to describe an mRNA that encodes only two proteins.

mRNA Circularization

In eukaryotes it is thought that mRNA molecules form circular structures due to an interaction between the cap binding complex and poly(A)-binding protein. Circularization is thought to promote recycling of ribosomes on the same message leading to efficient translation.

Degradation

Different mRNAs within the same cell have distinct lifetimes (stabilities). In bacterial cells, individual mRNAs can survive from seconds to more than an hour; in mammalian cells, mRNA lifetimes range from several minutes to days. The greater the stability of an mRNA the more protein may be produced from that mRNA. The limited lifetime of mRNA enables a cell to alter protein synthesis rapidly in response to its changing needs. There are many mechanisms that lead to the destruction of a mRNA, some of which are described below.

Prokaryotic mRNA Degradation

In general, in prokaryotes the lifetime of mRNA is much shorter than in eukaryotes. Prokaryotes degrade messages by using a combination of ribonucleases, including endonucleases, 3' exonucleases, and 5' exonucleases. In some instances, small RNA molecules (sRNA) tens to hundreds of nucleotides long can stimulate the degradation of specific mRNAs by base-pairing with complementary sequences and facilitating ribonuclease cleavage. It was recently shown that bacteria also have a sort of 5' cap consisting of a triphosphate on the 5' end. Removal of two of the phosphates leaves a 5' monophosphate, causing the message to be destroyed by the endonuclease RNase E.

Eukaryotic mRNA Turnover

Inside eukaryotic cells, there is a balance between the processes of translation and mRNA decay. Messages that are being actively translated are bound by ribosomes, the eukaryotic initiation factors eIF-4E and eIF-4G, and poly(A)-binding protein. eIF-4E and eIF-4G block the decapping enzyme (DCP2), and poly(A)-binding protein blocks the exosome complex, protecting the ends of the message. The balance between translation and decay is reflected in the size and abundance of cytoplasmic structures known as P-bodies The poly(A) tail of the mRNA is shortened by specialized exonucleases that are targeted to specific messenger RNAs by a combination of cis-regulatory sequences on the RNA and trans-acting RNA-binding proteins. Poly(A) tail removal is thought to disrupt the circular structure of the message and destabilize the cap binding complex. The message is then subject to degradation by either the exosome complex or the decapping complex. In this way, translationally inactive messages can be destroyed quickly, while active messages remain intact. The mechanism by which translation stops and the message is handed-off to decay complexes is not understood in detail.

AU-rich Element Decay

The presence of AU-rich elements in some mammalian mRNAs tends to destabilize those transcripts through the action of cellular proteins that bind these sequences and stimulate poly(A) tail removal. Loss of the poly(A) tail is thought to promote mRNA degradation by facilitating attack by both the exosome complex and the decapping complex. Rapid mRNA degradation via AU-rich elements is a critical mechanism for preventing the overproduction of potent cytokines such as tumor necrosis factor (TNF) and granulocyte-macrophage colony stimulating factor (GM-CSF). AU-rich elements also regulate the biosynthesis of proto-oncogenic transcription factors like c-Jun and c-Fos.

Nonsense Mediated Decay

Nonsense-mediated decay (NMD) is a cellular mechanism of mRNA surveillance that functions to detect nonsense mutations and prevent the expression of truncated or erroneous proteins. Following transcription, precursor mRNA undergoes an assemblage of ribonucleoprotein (RNP) components followed by regulatory pre-mRNA processing. Large average intron size in eukaryotic cells greatly increases the probability that aberrant mRNA splicing will result in the presence of a nonsense (stop) codon (UAA, UAG, UGA) somewhere within the open reading frame. NMD is triggered by exon junction complexes (EJCs) (components of the assembled RNP) that are deposited during pre-mRNA processing. While EJCs located in an open reading frame upstream of exon-exon junctions may facilitate ribosomal recruitment prior to displacement by a "pioneer" round of translation, EJCs located downstream of a nonsense codon are not displaced because the ribosome is released from the transcript before reaching it. These remaining EJCs function as tags for recruitment of UPF1 following the mRNA's transport out of the nucleus and into the cytosol where the RNA is degraded, for example by the exosome complex. NMD is not only a mechanism for degrading aberrant mRNA, however, as there are numerous examples of normal transcripts whose expression is regulated by this process including the plasticity protein Arc/Arg3.1. Eukaryotic messages are subject to surveillance by nonsense mediated decay (NMD), which checks for the presence of premature stop codons (nonsense codons) in the message. These can arise via incomplete splicing, V(D)J recombination in the adaptive immune system, mutations in DNA, transcription errors, leaky scanning by the ribosome causing a frame shift, and other causes. Detection of a premature stop codon triggers mRNA degradation by 5' decapping, 3' poly(A) tail removal, or endonucleolytic cleavage.

Small Interfering RNA (siRNA)

Small interfering RNA (siRNA), sometimes known as short interfering RNA or silencing RNA, is a class of double-stranded RNA molecules, 20-25 nucleotides in length, that play a variety of roles in biology. The most notable role of siRNA is its involvement in the RNA interference (RNAi) pathway, where it interferes with the expression of a specific gene. In addition to its role in the RNAi pathway, siRNA also acts in RNAi-related pathways, e.g., as an antiviral mechanism or in shaping the chromatin structure of a genome; the complexity of these pathways is only now being elucidated.

siRNAs were first discovered by David Baulcombe's group at the Sainsbury Laboratory in Norwich, England, as part of post-transcriptional gene silencing (PTGS) in plants. The group published their findings in Science in a 1999 paper titled "A species of small antisense RNA in posttranscriptional gene silencing in plants". Shortly thereafter, in 2001, synthetic siRNAs were shown to be able to induce RNAi in mammalian cells by Thomas Tuschl, and colleagues in a paper published in Nature. This discovery led to a surge in interest in harnessing RNAi for biomedical research and drug development.

In metazoans, small interfering RNAs (siRNAs) processed by Dicer are incorporated into a complex known as the RNA-induced silencing complex or RISC. This complex contains an endonuclease that cleaves perfectly complementary messages to which the siRNA binds. The resulting mRNA fragments are then destroyed by exonucleases. siRNA is commonly used in laboratories to block the function of genes in cell culture. It is thought to be part of the innate immune system as a defense against double-stranded RNA viruses. Structure siRNAs have a well-defined structure: a short (usually 21-nt) double-strand RNA (dsRNA) with 2-nt 3' overhangs on either end:

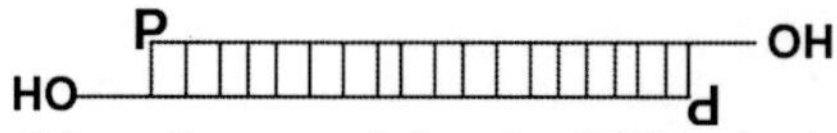

Schematic representation of a siRNA molecule: a ~19-21 basepair RNA core duplex that is followed by a 2 nucelotide 3' overhang on each strand. OH: 3' hydroxyl; P: 5' Phosphate

Each strand has a 5' phosphate group and a 3' hydroxyl (-OH) group. This structure is the result of processing by dicer, an enzyme that converts either long dsRNAs or small hairpin RNAs into siRNAs. siRNAs can also be exogenously (artificially) introduced into cells by various transfection methods to bring about the specific knockdown of a gene of interest. In essence, any gene whose sequence is known can, thus, be targeted based on sequence complementarity with an appropriately tailored siRNA. This has made siRNAs an important tool for gene function and drug target validation studies in the post-genomic era.

RNA Activation

It has recently been found that dsRNA can also activate gene expression, a mechanism that has been termed "small RNA-induced gene activation" or RNAa. It has been shown that dsRNAs targeting gene promoters induce potent transcriptional activation of associated genes. RNAa was demonstrated in human cells using synthetic dsRNAs, termed "small activating RNAs" (saRNAs). It is currently not known whether RNAa is conserved in other organisms.

Challenges: Avoiding Nonspecific Effects

Because RNAi intersects with a number of other pathways, it is not surprising that on occasion nonspecific effects are triggered by the experimental introduction of an siRNA. When a mammalian cell encounters a double-stranded RNA such as an siRNA, it may mistake it as a viral by-product and mount an immune response. Furthermore, because structurally related microRNAs modulate gene expression largely via incomplete complementarity base pair interactions with a target mRNA, the introduction of an siRNA may cause unintended off-targeting.

Innate Immunity

Introduction of too much siRNA can result in nonspecific events due to activation of innate immune responses. Most evidence to date suggests that this is probably due to activation of the dsRNA sensor PKR, although retinoic acid-inducible gene I (RIG-I) may also be involved· The induction of cytokines via toll-like receptor 7 (TLR7) has also been described. One promising method of reducing the nonspecific effects is to convert the siRNA into a microRNA. MicroRNAs occur naturally, and by harnessing this endogenous pathway it should be possible to achieve similar gene knockdown at comparatively low concentrations of resulting siRNAs. This should minimize nonspecific effects.

Off-targeting

Off-targeting is another challenge to the use of siRNAs as a gene knockdown tool. Here, genes with incomplete complementarity are inadvertently downregulated by the siRNA (in effect, the siRNA acts as a miRNA), leading to problems in data interpretation and potential toxicity.

This, however, can be partly addressed by designing appropriate control experiments, and siRNA design algorithms are currently being developed to produce siRNAs free from off-targeting. Genome-wide expression analysis, e.g., by microarray technology, can then be used to verify this and further refine the algorithms. A 2006 paper from the laboratory of Dr. Khvorova implicates 6- or 7-basepair-long stretches from position 2 onward in the siRNA matching with 3'UTR regions in off-targeted genes.

The RNA-induced Silencing Complex

The mature miRNA is part of an active RNA-induced silencing complex (RISC) containing Dicer and many associated proteins. RISC is also known as a microRNA ribonucleoprotein complex (miRNP); RISC with incorporated miRNA is sometimes referred to as "miRISC."

Dicer processing of the pre-miRNA is thought to be coupled with unwinding of the duplex. Generally, only one strand is incorporated into the miRISC, selected on the basis of its thermodynamic instability and weaker base-pairing relative to the other strand.

The position of the stem-loop may also influence strand choice. The other strand, called the passenger strand due to its lower levels in the steady state, is denoted with an asterisk (*) and is normally degraded. In some cases, both strands of the duplex are viable and become functional miRNA that target different mRNA populations.

Members of the argonaute (Ago) protein family are central to RISC function. Argonautes are needed for miRNA-induced silencing and contain two conserved RNA binding domains: a PAZ domain that can bind the single stranded 3' end of the mature miRNA and a PIWI domain that structurally resembles ribonuclease-H and functions to interact with the 5' end of the guide strand. They bind the mature miRNA and orient it for interaction with a target mRNA. Some argonautes, for example human Ago2, cleave target transcripts directly;

argonautes may also recruit additional proteins to achieve translational repression. The human genome encodes eight argonaute proteins divided by sequence similarities into two families: AGO (with four members present in all mammalian cells and called E1F2C/hAgo in humans), and PIWI (found in the germ line and hematopoietic stem cells).

Additional RISC components include TRBP [human immunodeficiency virus (HIV) transactivating response RNA (TAR) binding protein], PACT (protein activator of the interferon induced protein kinase (PACT), the SMN complex, fragile X mental retardation protein (FMRP), and Tudor staphylococcal nuclease-domain-containing protein (Tudor-SN). Strands

The RNA endonuclease Dicer plays a role in aiding RISC action by providing the initial RNA material to activate the complex as well as the first RNA substrate molecule.

When the Dicer, which has endonuclease activity against dsRNA and pre-miRNAs, cleaves a pre-miRNA stem-loop or a dsRNA, a 20- to 25-base-pair double-stranded RNA fragment is formed with a two-nucleotide 3' overhang at each end.

- One strand is integrated into the RISC complex. This strand is known as the guide strand and is selected by the argonaute protein, the catalytically active RNase in the RISC complex, on the basis of the thermodynamic stability of the 5' end. The strand with a less thermodynamically stable 5' end is loaded into the RISC complex.
- The remaining strand, known as the anti-guide or passenger strand, is degraded as a RISC complex substrate.

Binding

The RISC complex with a bound siRNA recognizes complementary messenger RNA (mRNA) molecules and degrades them, resulting in substantially decreased levels of protein translation and effectively turning off the gene.

It is as yet unclear how the activated RISC complex locates the mRNA targets in the cell, though it has been shown that the process can occur in situations outside of ongoing protein translation from mRNA. Endogenously expressed miRNA in metazoans is usually imperfectly complementary to a large number of nuclear genes and has a modulating effect on these genes' levels of expression via translational repression.

However, in plants the process has a much greater specificity to the mRNA and usually each miRNA only binds to one mRNA. Recently identified members of RISC complex are SND1(p100 Coactivator or Tudor SN)and AEG-1(MTDH). AEG-1 and SND1 are oncogenes and regulates various gene expression.

Mode of Silencing

Gene silencing may occur either via mRNA degradation or preventing mRNA from being translated. It has been demonstrated that if there is complete complementation between the miRNA and target mRNA sequence, Ago2 can cleave the mRNA and lead to direct mRNA degradation. Yet, if there isn't complete complementation the silencing is achieved by preventing translation.

miRNA Turnover

Turnover of mature miRNA is needed for rapid changes in miRNA expression profiles. During miRNA maturation in the cytoplasm, uptake by the Argonaute protein is thought to stabilize the guide strand, while the opposite (* or "passenger") strand is preferentially destroyed. In what has been called a "Use it or lose it" strategy, Argonaute may preferentially retain miRNAs with many targets over miRNAs with few or no targets, leading to degradation of the non-targeting molecules.

Decay of mature miRNAs in Caenorhabditis elegans is mediated by the 5′-to-3′ exoribonuclease XRN2, also known as Rat1p. In plants, SDN (small RNA degrading nuclease) family members degrade miRNAs in the opposite (3'-to-5') direction. Similar enzymes are encoded in animal genomes, but their roles have not yet been described.

Several miRNA modifications affect miRNA stability. As indicated by work in the model organism Arabidopsis thaliana (thale cress), mature plant miRNAs appear to be stabilized by the addition of methyl moieties at the 3' end. The 2'-O-conjugated methyl groups block the addition of uracil (U) residues by uridyltransferase enzymes, a modification that may be associated with miRNA degradation. However, uridylation may also protect some miRNAs; the consequences of this modification are incompletely understood. Uridylation of some animal miRNAs has also been reported. Both plant and animal miRNAs may be altered by addition of adenine (A) residues to the 3' end of the miRNA. An extra A added to the end of mammalian miR-122, a liver-enriched miRNA important in Hepatitis C, stabilizes the molecule, and plant miRNAs ending with an adenine residue have slower decay rates.

Cellular Functions

The function of miRNAs appears to be in gene regulation. For that purpose, a miRNA is complementary to a part of one or more messenger RNAs (mRNAs). Animal miRNAs are usually complementary to a site in the 3' UTR whereas plant miRNAs are usually complementary to coding regions of mRNAs. Perfect or near perfect base pairing with the target RNA promotes cleavage of the RNA. This is the primary mode of plant miRNAs. In animals miRNAs more often have only partly the right sequence of nucleotides to bond with the target mRNA. The match-ups are imperfect. Animal miRNAs inhibit protein translation of the target mRNA (this exists in plants as well but is less common). MicroRNAs that are partially complementary to a target can also speed up deadenylation, causing mRNAs to be degraded sooner. For partially complementary microRNAs to recognise their targets, nucleotides 2–7 of the miRNA (its 'seed region') still have to be perfectly complementary. miRNAs occasionally also cause histone modification and DNA methylation of promoter sites, which affects the expression of target genes.

Unlike plant microRNAs, the animal microRNAs target a diverse set of genes. However, genes involved in functions common to all cells, such as gene expression, have relatively fewer microRNA target sites and seem to be under selection to avoid targeting by microRNAs.

dsRNA can also activate gene expression, a mechanism that has been termed "small RNA-induced gene activation" or RNAa. dsRNAs targeting gene promoters can induce potent transcriptional activation of associated genes. This was demonstrated in human cells us as also been demonstrated for endogenous microRNA. Interactions between microRNAs and complementary sequences on genes and even pseudogenes that share sequence homology are thought to be a back channel of communication regulating expression levels between paralogous genes. Given the name "competing endogenous RNAs" (ceRNAs), these microRNAs bind to "microRNA response elements" on genes and pseudogenes and may provide another explanation for the persistence of non-coding ("junk") DNA.

Evolution

MicroRNAs are significant phylogenetic markers because of their astonishingly low rate of evolution. Their origin may have permitted the development of morphological innovation, and by making gene expression more specific and 'fine-tunable', permitted the genesis of complex organs and perhaps, ultimately, complex life. Indeed, rapid bursts of morphological innovation are generally associated with a high rate of microRNA accumulation.

MicroRNAs originate predominantly by the random formation of hairpins in "non-coding" sections of DNA (i.e. introns or intergene regions), but also by the duplication and modification of existing microRNAs. The rate of evolution (i.e. nucleotide substitution) in recently-originated microRNAs is comparable to that elsewhere in the non-coding DNA, implying evolution by neutral drift; however, older microRNAs have a much lower rate of change (often less than one substitution per hundred million years), suggesting that once a microRNA gains a function it undergoes extreme purifying selection. At this point, a microRNA is rarely lost from an animal's genome, although microRNAs which are more recently derived (and thus presumably non-functional) are frequently lost. This makes them a valuable phylogenetic marker, and they are being looked upon as a possible solution to such outstanding phylogenetic problems as the relationships of arthropods.

MicroRNAs feature in the genomes of most eukaryotic organisms, from the brown algae to the metazoa. Across all species, in excess of 5000 had been identified by March 2010. Whilst short RNA sequences (50 – hundreds of base pairs) of a broadly comparable function occur in bacteria, bacteria lack true microRNAs.

Experimental Detection and Manipulation of miRNA

MicroRNA expression can be quantified in a two-step polymerase chain reaction process of modified RT-PCR followed by quantitative real-time PCR. Variations of this method achieve absolute or relative quantification. miRNAs can also be hybridized to microarrays, slides or chips with probes to hundreds or thousands of miRNA targets, so that relative levels of miRNAs can be determined in different samples. MicroRNAs can be both discovered and profiled by high-throughput sequencing methods.

The activity of an miRNA can be experimentally inhibited using a locked nucleic acid (LNA) oligo, a Morpholino oligo or a 2'-O-methyl RNA oligo. Additionally, a specific miRNA can be silenced by a complementary antagomir. MicroRNA maturation can be inhibited at several points by steric-blocking oligos. The miRNA target site of an mRNA transcript can also be blocked by a steric-blocking oligo. For the "in situ" detection of miRNA, the use of LNA is currently the only efficient method. The locked conformation of LNA results in enhanced hybridization properties and increases sensitivity and selectivity, making it ideal for detection of short miRNA.

miRNA and Disease

Just as miRNA is involved in the normal functioning of eukaryotic cells, so has dysregulation of miRNA been associated with disease. A manually curated, publicly available database miR2Disease documents known relationships between miRNA dysregulation and human disease.

miRNA and Inherited Diseases

A mutation in the seed region of miR-96 causes hereditary progressive hearing loss. A mutation in the seed region of miR-184 causes hereditary keratoconus with anterior polar cataract. Deletion of the miR-17~92 cluster causes skeletal and growth defects.

miRNA and Cancer

Several miRNAs have been found to have links with some types of cancer. MicroRNA-21 is one of the first microRNAs that was identified as an oncomiR.

A study of mice altered to produce excess c-Myc — a protein with mutated forms implicated in several cancers — shows that miRNA has an effect on the development of cancer. Mice that were engineered to produce a surplus of types of miRNA found in lymphoma cells developed the disease within 50 days and died two weeks later.

In contrast, mice without the surplus miRNA lived over 100 days. Leukemia can be caused by the insertion of a viral genome next to the 17-92 array of microRNAs leading to increased expression of this microRNA.

Another study found that two types of miRNA inhibit the E2F1 protein, which regulates cell proliferation. miRNA appears to bind to messenger RNA before it can be translated to proteins that switch genes on and off.

By measuring activity among 217 genes encoding miRNA, patterns of gene activity that can distinguish types of cancers can be discerned. miRNA signatures may enable classification of cancer. This will allow doctors to determine the original tissue type which spawned a cancer and to be able to target a treatment course based on the original tissue type. miRNA profiling has already been able to determine whether patients with chronic lymphocytic leukemia had slow growing or aggressive forms of the cancer.

Transgenic mice that over-express or lack specific miRNAs have provided insight into the role of small RNAs in various malignancies.

A novel miRNA-profiling based screening assay for the detection of early-stage colorectal cancer has been developed and is currently in clinical trials. Early results showed that blood plasma samples collected from patients with early, resectable (Stage II) colorectal cancer could be distinguished from those of sex-and age-matched healthy volunteers. Sufficient selectivity and specificity could be achieved using small (less than 1 mL) samples of blood. The test has potential to be a cost-effective, non-invasive way to identify at-risk patients who should undergo colonoscopy.

Another role for miRNA in cancers is to use their expression level as a prognostic, for example one study on NSCLC samples found that low miR-324a levels could serve as a prognostic indicator of poor survival, another found that either high miR-185 or low miR-133b levels correlated with metastasis and poor survival in colorectal cancer.

miRNA and Heart Disease

The global role of miRNA function in the heart has been addressed by conditionally inhibiting miRNA maturation in the murine heart, and has revealed that miRNAs play an essential role during its development. miRNA expression profiling studies demonstrate that expression levels of specific miRNAs change in diseased human hearts, pointing to their involvement in cardiomyopathies. Furthermore, studies on specific miRNAs in animal models have identified distinct roles for miRNAs both during heart development and under pathological conditions, including the regulation of key factors important for cardiogenesis, the hypertrophic growth response, and cardiac conductance.

miRNA and the Nervous System

miRNAs appear to regulate the nervous system. Neural miRNAs are involved at various stages of synaptic development, including dendritogenesis (involving miR-132, miR-134 and miR-124), synapse formation and synapse maturation (where miR-134 and miR-138 are thought to be involved). Some studies find altered miRNA expression in schizophrenia.

miRNA and Non-coding RNAs

When the human genome project mapped its first chromosome in 1999, it was predicted the genome would contain over 100,000 protein coding genes. However, only around 20,000 were eventually identified (International Human Genome Sequencing Consortium, 2004). Since then, the advent of bioinformatics approaches combined with genome tiling studies examining the transcriptome, systematic sequencing of full length cDNA libraries, and experimental validation (including the creation of miRNA derived antisense oligonucleotides called antagomirs) have revealed that many transcripts are non protein-coding RNA, including several snoRNAs and miRNAs.

Other Decay Mechanisms

There are other ways by which messages can be degraded, including non-stop decay and silencing by Piwi-interacting RNA (piRNA), among others.

Untranslated Region (UTR)

In molecular genetics, an untranslated region (or UTR) refers to either of two sections on each side of a coding sequence on a strand of mRNA. If it is found on the 5' side, it is called the 5' UTR (or leader sequence), or if it is found on the 3' side, it is called the 3' UTR (or trailer sequence).

Primary Transcript

A primary transcript is an RNA molecule that has not yet undergone any modification after its synthesis. For example, a precursor messenger RNA (pre-mRNA) is a primary transcript that becomes a messenger RNA (mRNA) after processing, and a primary microRNA (pri-miRNA) precursor becomes a microRNA (miRNA) after processing.

Spliceosome

A spliceosome is a complex of snRNA and protein subunits that removes introns from a transcribed pre-mRNA (hnRNA) segment. This process is generally referred to as splicing.

Composition

Each spliceosome is composed of five small nuclear RNAs (snRNAs), and a range of associated protein factors. The snRNAs that make up the nuclear spliceosome are named U1, U2, U4, U5, and U6, and participate in several RNA-RNA and RNA-protein interactions. The RNA component of the small nuclear ribonucleic protein or snRNP (pronounced "snurp") is rich in uridine (the nucleoside analog of the uracil nucleotide). The canonical assembly of the spliceosome occurs anew on each hnRNA (pre-mRNA). The hnRNA contains specific sequence elements that are recognized and utilized during spliceosome assembly. These include the 5' end splice, the branch point sequence, the polypyrimidine tract, and the 3' end splice site. The spliceosome catalyzes the removal of introns, and the ligation of the flanking exons. Introns typically have a GU nucleotide sequence at the 5' end splice site, and an AG at the 3' end splice site. The 3' splice site can be further defined by a variable length of polypyrimidines, called the polypyrimidine tract (PPT), which serves the dual function of recruiting factors to the 3' splice site and possibly recruiting factors to the branch point sequence (BPS). The BPS contains the conserved Adenosine required for the first step of splicing. A group of less abundant snRNAs, U11, U12, U4atac, and U6atac, together with U5, are subunits of the so-called minor spliceosome that splices a rare class of pre-mRNA introns, denoted U12-type. These snRNPs form the U12 spliceosome are located in the cytosol. New evidence derived from the first crystal structure of a group II intron suggests that the spliceosome is actually a ribozyme, and that it uses a two–metal ion mechanism for catalysis.

Alternative Splicing

Alternative splicing (the re-combination of different exons) is a major source of genetic diversity in eukaryotes. Splice variants have been used to account for the relatively small number of genes in the human genome. For years the estimate widely varied with top estimates reaching 100,000 genes, but now, due to the Human Genome Project the figure is believed to be closer to 20,000 genes. One particular Drosophila gene (Dscam, the Drosophila homolog of the human Down syndrome cell adhesion molecule DSCAM) can be alternatively spliced into 38,000 different mRNA.

RNA Splicing

In molecular biology and genetics, splicing is a modification of an RNA after transcription, in which introns are removed and exons are joined. This is needed for the typical eukaryotic messenger RNA before it can be used to produce a correct protein through translation. For many eukaryotic introns, splicing is done in a series of reactions which are catalyzed by the spliceosome, a complex of small nuclear ribonucleoproteins (snRNPs), but there are also self-splicing introns. In 1977, work by the Sharp and Roberts labs revealed that genes of higher organisms are "split" or present in several distinct segments along the DNA molecule. The coding regions of the gene are separated by non-coding DNA that is not involved in protein expression.

The split gene structure was found when adenoviral mRNAs were hybridized to endonuclease cleavage fragments of single stranded viral DNA. It was observed that the mRNAs of the mRNA-DNA hybrids contained 5' and 3' tails of non-hydrogen bonded regions. When larger fragments of viral DNAs were used, forked structures of looped out DNA were observed when hybridized to the viral mRNAs. It was realized that the looped out regions, the introns, are excised from the precursor mRNAs in a process Sharp named "splicing".

The split gene structure was subsequently found to be common to most eukaryotic genes. Phillip Sharp and Richard J. Roberts were awarded the 1993 Nobel Prize in Physiology or Medicine for their discovery of introns and the splicing process.

Spliceosome Assembly

The model for formation of the spliceosome active site involves an ordered, stepwise assembly of discrete snRNP particles on the hnRNA substrate. The first recognition of hnRNAs involves U1 snRNP binding to the 5' end splice site of the hnRNA and other non-snRNP associated factors to form the commitment complex, or early (E) complex in mammals.

The commitment complex is an ATP-independent complex that commits the hnRNA to the splicing pathway. U2 snRNP is recruited to the branch region through interactions with the E complex component U2AF (U2 snRNP auxiliary factor) and possibly U1 snRNP. In an ATP-dependent reaction, U2 snRNP becomes tightly associated with the branch point sequence (BPS) to form complex A. A duplex formed between U2 snRNP and the hnRNA branch region bulges out the branch adenosine specifying it as the nucleophile for the first transesterification.

The presence of a pseudouridine residue in U2 snRNA, nearly opposite of the branch site, results in an altered conformation of the RNA-RNA duplex upon the U2 snRNP binding. Specifically, the altered structure of the duplex induced by the pseudouridine places the 2' OH of the bulged adenosine in a favorable position for the first step of splicing.

The U4/U5/U6 tri-snRNP is recruited to the assembling spliceosome to form complex B, and following several rearrangements, complex C (the spliceosome) is activated for catalysis. It is unclear how the triple snRNP is recruited to complex A, but this process may be mediated through protein-protein interactions and/or base pairing interactions between U2 snRNA and U6 snRNA.

The U5 snRNP interacts with sequences at the 5' and 3' splice sites via the invariant loop of U5 snRNA and U5 protein components interact with the 3' splice site region.

Upon recruitment of the triple snRNP, several RNA-RNA rearrangements precede the first catalytic step and further rearrangements occur in the catalytically active spliceosome. Several of the RNA-RNA interactions are mutually exclusive; however, it is not known what triggers these interactions, nor the order of these rearrangements.

The first rearrangement is probably the displacement of U1 snRNP from the 5' splice site and formation of a U6 snRNA interaction. It is known that U1 snRNP is only weakly associated with fully formed spliceosomes, and U1 snRNP is inhibitory to the formation of a U6-5' splice site interaction on a model of substrate oligonucleotide containing a short 5' exon and 5' splice site. Binding of U2 snRNP to the branch point sequence (BPS) is one example of an RNA-RNA interaction displacing a protein-RNA interaction. Upon recruitment of U2 snRNP, the branch binding protein SF1 in the commitment complex is displaced since the binding site of U2 snRNA and SF1 are mutually exclusive events.

Within the U2 snRNA, there are other mutually exclusive rearrangements that occur between competing conformations. For example, in the active form, stem loop IIa is favored; in the inactive form a mutually exclusive interaction between the loop and a downstream sequence predominates.

It is unclear how U4 is displaced from U6 snRNAm, although RNA has been implicated in spliceosome assembly, and may function to unwind U4/U6 and promote the formation of a U2/U6 snRNA interaction. The interactions of U4/U6 stem loops I and II dissociate and the freed stem loop II region of U6 folds on itself to form an intramolecular stem loop and U4 is no longer required in further spliceosome assembly. The freed stem loop I region of U6 base pairs with U2 snRNA forming the U2/U6 helix I. However, the helix I structure is mutually exclusive with the 3' half of an internal 5' stem loop region of U2 snRNA.

Bibliography

Ambrose, E. J., and Easty, D. M.: *Cell Biology: Reading,* Addison- Wesley, Mass, 1970.

Barrington, E. J. W.: *Biochemistry of Primitive Deuterostomians*, London, Academic Press, 1974.

Bauer, M.W., and G. Gaskell : *Biotechnology: The Making of a Global Controversy*, Cambridge: Cambridge University Press, 2002.

Bowler, Peter J.: *The Mendelian Revolution: The Emergence of Hereditarian Concepts in Modern Science and Society*. Baltimore: Johns Hopkins University Press, 1989.

Broach, J.R. : *The Molecular Biology of the Yeast,* Cold Spring Harbor: Cold Spring Harbor Laboratory, 1981.

Carroll SB, Grenier J, Weatherbee SD: *From DNA to Diversity: Molecular Genetics and the Evolution of Animal Design*. Oxford: Blackwell Publishing, 2005.

Chauhan, B.S. : *Principles of Biochemistry and Biophysics,* Laxmi Publications, Delhi, 2008.

Dhruve, J J : *Biochemistry at a Glance*, Agrotech Pub, Delhi, 2007.

Friedberg, E.C. : *DNA Repair,* New York: WH Freeman and Company, 1985.

Ganguly, Smriti : *Biochemistry of Biomolecules,* Pearl Books, Delhi, 2007.

Harris, John : *Wonderwoman and Superman: The Ethics of Human Biotechnology,* Oxford: Oxford University Press, 1992.

Hartl D, Jones E: *Genetics: Analysis of Genes and Genomes*. Jones & Bartlett, 2005.

Kitcher, Philip : *The Lives to Come: The Genetic Revolution and Human Possibilities*, New York: Simon and Schuster, 1996.

Lata Bhattacharya: *Biochemistry of Nutrition*, Discovery, Delhi, 2010.

Lodish, Harvey : *Molecular Cell Biology*, New York: W. H. Freeman, 2000.

Morholt, E. and Brandwein, P.F. : *Sourcebook for the Biological Sciences,* San Diego: Harcourt Brace Jovanovich, 1986.

Morris, Jonathan : *Ethics and Biotechnology*, New York: Chelsea House, 2005.

Narvekar, Raghunath : *Molecular Biochemistry : Principles and Practices,* Adhyayan Pub, Delhi, 2008.

Patel, S V ; U K Kandoliya: *Biochemistry at a Glance*, Agrotech Pub, Delhi, 2007.

Primrose, S. B. : *Molecular Biotechnology,* Boston: Blackwell Scientific Publications, 1991.

Reuter, Lars : *Modern Biotechnology in Postmodern Times? A Reflection on European Policies and Human Agency,* Dordrecht, Netherlands: Springer, 2003.

Sarath Chandra Bose: *Biochemistry : A Practical Manual*, Pharma Med Press, Delhi, 2010.

Sharma, Pradeep: *Biochemistry and Organisation of Cells*, RBSA Pub, Delhi, 2006.

Singh, Bimla: *Biochemistry and Genomic Revolution*, Vista International Pub, Delhi, 2006.

Trevor H. : *Transforming Matter: A History of Chemistry from Alchemy to the Buckyball,* Baltimore: Johns Hopkins University Press, 2001.

William Alec : *The Chemical Industry*, London: Longman Group Limited, 1971.

Yadav, Neelam: *Biochemistry and Biotechnology : A Laboratory Manual*, Pointer, Delhi, 2007.

Index

❑❑❑